I0476099

DPs Dominion

Publishing Services

http://www.dominionpublishingstores.yolasite.com

Agricultural Engineering
Principles & Practice

Volume 2

Segun R. Bello

B. Eng (Hons), FUT, Akure, MSc, Ibadan,
MNSE, MNIAE, FSINRHD, R. Engr. (COREN)

AGRICULTURAL ENGINEERING
Principles & Practice

Volume 2 2nd Edition

ISBN-13: 978- 145-633-568-7
ISBN: 145-633-568-5

First published in September 2012

Printed by Createspace US
7290 Investment Drive
Suite B North Charleston,
SC 29418 USA, www.createspace.com

List of Contributors

Dr. T. A. Adegbulugbe (PhD, Ibadan)

Former Head, Department of Agricultural Engineering Federal College of Agriculture, Moore plantation, Ibadan

Engr. David Aremu (MSc, Ibadan)

Department of Agricultural Engineering Federal College of Agriculture, Moore Plantation, Ibadan

Dedication

To the glory of God Almighty

Acknowledgement

Unlimited gratitude goes to God Almighty, the author of life and the giver of knowledge, for His grace and inspirations in the pursuit of this divine agenda in the course of my career. Glory be to His name.

The author deeply appreciate the following contributors; Dr. T. A. Adegbulugbe, and Engr. David Aremu, both of Federal College of Agriculture Moor plantation Ibadan, Engr. Mohammed A. Suleiman of the Department of Agriculture and Bio-Environmental, College of Agriculture Jalingo and the several teachers, authors and researchers whose wealth of experiences documented in books, journal and print forms, as well as numerous web materials on new innovations and development in agricultural engineering help package this work.

I sincerely thank all students, past and present, of the departments of agricultural technology, agricultural engineering and engineering technology, Federal College of Agriculture Ishiagu, Federal College of Agriculture Moor plantation Ibadan, College of Agriculture Jalingo and Michael Okpara University of Agriculture, Umudike and all who had come in contact with my books in various fields of agricultural engineering practice. Your feed-backs and valuable comments and criticisms have been used to review and upgrade this work packaged in 2 volumes.

Despite all the help received from many people, it seems inevitable that there will be some inaccuracies or errors in the text. For these the author accepts responsibility and apologizes in advance for any incorrect statements or impressions given. Should errors be noticed, the author would welcome factual corrections. He would also be happy to receive comments, observations and additional information on any topic, section or statements in any part of the book. This would be particularly useful should any updated or translated edition be planned. Correspondence may be addressed to the author

My special thanks go to my dear friend, companion and wife, who had always back-up the realization of God's plan for me. She is a virtuous woman and help meet indeed. Her understanding and tolerance in taking full responsibility of running our home during the entire review and upgrade exercise are quite commendable.

I am grateful to my children, Ayomikun, Pelumi, Damilola and Adeola, who were so wonderful and cooperative during this period. I am greatly encouraged and strengthened by their prayers, my God shall surely reward them. Amen

Content

Preface to Volume 2

Agricultural engineering principles and practice Volume 2 is the continuation of exposition on the principles and practice of agricultural engineering. Volume 1 provide relevant background information at all levels of engineering training, professionalisms, career development, and in the practice of agricultural engineering undertakings spanning field preparation, crop planting and harvesting etc.

In this volume, soil and water resources, crop processing methods, farm structures, livestock housing, animal thermal environment and equipment were discussed in 3 parts.

Part 1 discussed soils and water conservation practice essential for crop establishment. Basic principles of soil and water resources management, erosion control, drainage requirement, dam design and management. Other topics include plant water requirements for irrigation, basic design procedures for all types of irrigation, drainage and flood control measures.

Part 2: The rheological properties that influence the behaviour of agricultural materials during technological processes were packaged in 5 chapters of this book. Crop harvest systems, agricultural postharvest systems, crop harvest and processing equipment, fruit processing and storage facilities were common features in this part.

Part 3 laid credence to structural requirements for man, animal and crop products on the farm as presented in 3 chapters, farmstead planning, farmstead resources and layout plan, farm storage structures, animal housing and thermal requirement for housing and processing of agricultural products.

The scope of agricultural engineering practice is inexhaustible and that informs a continual development and expansion of knowledge as advancements takes place.

Bello RS

480001, Nigeria

Part 1

SOIL AND WATER RESOURCES

CHAPTER 1

Soil and Water Resources

1   Introduction

Soil and water are two natural resources essential to support plant growth and establishment. On one hand, soil is the medium on which plants grow, while it requires water to transport or distribute mineral resources within the soil matrix as well as through the plant micro structures. In order word soil serves as the support for plant while water is the medium through which plant nutrients gets to the plant and also exchange of ions within soil particles.

These resources are so vital to agricultural sustainability that their depletion could cause major setback to agricultural production. These resources are prone to degradation and are as well exhaustible hence they must be conserved. Soil and water conservation are therefore part of the complex engineering problems which has arose interest in the emergence of a branch of agricultural engineering known as soil and water conservation engineering. These resources; their development and conservation are discussed in the following sections.

1.1   Soil resources

The primary building materials of soil consist of mineral particles ranging in size from stones to sand and clay. Organic matter and clay bond the larger particles together and, in some soils, calcium or iron and aluminum compounds also act as bonding agents. A deposit of soil material, resulting from one or more geological processes is subjected to further physical and chemical changes which are brought about by climate and other subsequently prevalent factors like rainfall, temperature etc.

For instance, as rainfall begins, vegetation starts to develop and the processes of leaching and eluviation of the surface of the soil material continued, gradually, with the passage of time, profound changes take place in the character of the soil.

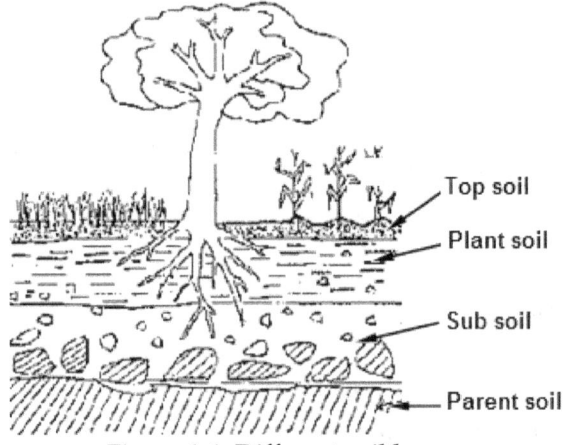

Figure 1-1: Different soil layers

These changes bring about the development of what is known as 'soil profile'.

Soil profile

Soil profile is described as a natural succession (arrangement) of zones or strata below the ground surface which represents the alterations in the original soil material which have been brought about by weathering processes. A fully developed soil such as is found in an undisturbed soil (forest) comprises of a mature layer of several millimeters or more of leaves and other under composed plant materials. Below this layer, there is a layer of decomposed organic materials (humus) and mineral material. A typical soil profile is shown in Figure 1-2.

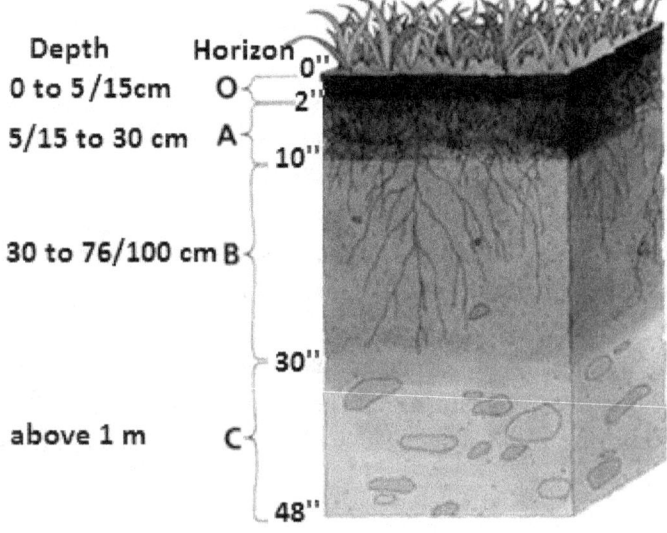

Figure 1-2: Typical soil profile (Wikipedia)

Generally, three distinct strata or horizons occur in a natural soil-profile. This number may increase to five or more in soils which are very old or in which the weathering processes have been unusually intense. From top to bottom these horizons are designated as the O-horizon, A-horizon, the B-horizon and the C-horizon. Each of these horizons may consist of sub-horizons with distinctive physical and chemical characteristics and may be designated as A_1, A_2, B_1, B_2, C_1, C_2 etc. The transition between horizons and sub-horizons may not be sharp but gradual. At a certain place, one or more horizons may be missing in the soil profile for special reasons.

Description of soil horizons

O-Horizon: The O-Horizon composed of dry organic plant residues at the topmost layer and decaying plant materials. This is a zone of active micro organism activities and ranges between 0 and 5 cm depth but could be up to 15 cm.

A-horizon: The A-horizon is rich in humus and organic plant residue. The horizon A or topsoil is the most fertile and in a downward direction successive horizon becomes less fertile.This is usually eluviated and leached; that is, the ultrafine colloidal material and the soluble mineral salts are washed out of this horizon by percolating water. It is dark in colour and its thickness may range from a few centimetres to half a metre. This horizon often exhibits many undesirable engineering characteristics and is of value only to agricultural soil scientists.

B-horizon: The B-horizon is sometimes referred to as the zone of accumulation. The material which has migrated from the A-horizon by leaching and eluviation gets deposited in this zone. There is a distinct difference of colour between this zone and the dark top soil of the A-horizon. This soil is very much chemically active at the surface and contains unstable fine-grained material. Thus, this is important in highway and airfield construction work and light structures such as single storey residential buildings, in which the foundations are located near the ground surface. The thickness of B-horizon may range from 0.50 to 0.75 m and up to 1m in some cases.

C-horizon: The material in the C-horizon is in the same physical and chemical state as it was first deposited by water, wind or ice in the geological cycle. The thickness of this horizon may range from a few centimetres to more than 3 m. The upper region of this horizon is often oxidised to a considerable extent. It is from this horizon that the bulk of the material is often borrowed for the construction of large soil structures such as earth dams.

Characteristics of each horizon

O-Horizon: Organic residues, dark humus

A-Horizon: Light brown loam, leached

B-Horizon: Dark brown clay, leached

C_1-Horizon: Light brown silty clay, oxidised and unleached

C_2-Horizon: Light brown silty clay, unoxidised and unleached

Soil structure

The shape and arrangement of soil particle is called *structures*. The 'structure' of a soil may be defined as the manner of arrangement and state of aggregation of soil grains. Structure describes the size, shape and stability of the solid soil material and the size, shape and continuity of the spaces (pores) between the soil solids. The strength of the bonds holding the solid particles together determines the stability of soil structure and its potential to withstand the effects of external forces. The stability of soil structure can change due to increases or decreases in the quantity or quality of binding agents such as organic matter.

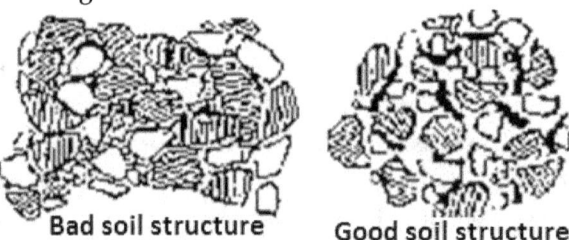

Bad soil structure   Good soil structure

Figure 1-3: Soil structure

Soil structure has been described as the "architecture" of the soil. In a broader sense, consideration of mineralogical composition, electrical properties, orientation and shape of soil grains, nature and properties of soil water and the interaction of soil water and soil grains, also may be included in the study of soil structure, which is typical for transported or sedimented soils.

Structural composition of sedimented soils influences, many of their important engineering properties such as permeability, compressibility and shear strength. The following types of structure are commonly studied:

1   *Single-grained structure*: Single-grained structure is characteristic of coarse-grained soils; with a particle size greater than 0.02 mm. Gravitational forces are higher than the surface forces and hence grain to grain contact results.

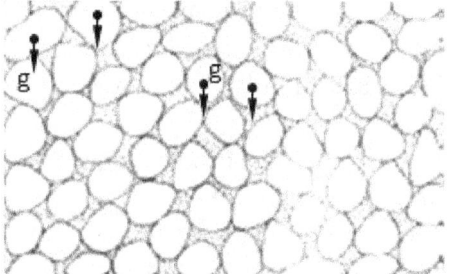

Figure 1-4: Single-grained structure

2   *Honey-comb structure*: This structure can occur only in fine-grained soils, especially in silt and rock flour. Due to the relatively smaller size of grains, besides gravitational forces, inter-particle surface forces also play an important role in the process of settling down. In the formation of a honey-comb structure, each cell of a honey-comb being made up of numerous individual soil grains. The structure has a large void space and may carry high loads without a significant volume change. The structure can be broken down by external disturbances.

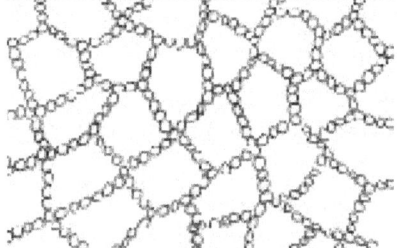

Figure 1-5: Honey-comb structure

3   *Flocculent structure*: This structure is characteristic of fine-grained soils such as clays. Inter-particle forces play a predominant role in the deposition.

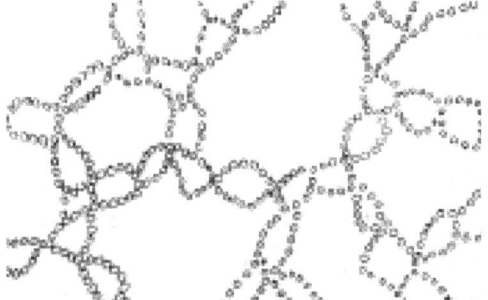

Figure 1-6: Flocculent structure

Soil structural deterioration

Under natural conditions, deterioration of soil structure is uncommon. It occurs along tracks used regularly by foraging animals, around drinking holes and where plant cover is removed, for example by fire. Where agriculture is practiced, changes in soil structure are common. They can be divided into two main categories:

1. Those associated with a net reduction in soil organic matter in the topsoil, and
2. Those changes resulting directly from the reaction of the soil to an applied force (e.g. tractor traffic).

Soil texture

The term *'texture'* refers to the appearance of the surface of soils. Texture of a soil is reflected largely by the particle size, shape, and gradation. The concept of texture of a soil has found some use in the classification of soils. Soil is generally classified into three: clay, silt and sand. Clay is the finest while sand is coarse.

Functions of soil

Soil is about the most important long-term resource base for the support of plant and animal life directly or indirectly. Generally, the development of any society is determined to a large extent by its capacity to explore the resources of the soil (biosphere). Soil resources serve the following functions

1. Soil provides physical support for plant growth, animals and engineering structures,
2. Soil serves as sources of food for plant and animals,
3. Soil also provides surface for such activities as grazing, mining and establishment of settlements.

1.2   Water resources

Water resources

Humans can truly be called water creatures: 65% of our body weight is water. We live on a planet made largely of water: 70% of the surface is water, with a volume of 330 million cubic meters, representing 10% of the planet's mass. The amount of usable freshwater actually comprises less than 1% of the earth's water. Of this 1%, 98% is groundwater and the remaining 2% is fresh surface water (Van Morrill and Gabrielle Belfit, 1999).

Agriculture uses and needs plenty of water, so also are the other sectors like the industry, municipals, recreation etc. As the demand for a share of the limited water supply of the nation increases it will become increasingly important for agriculture to improve the efficiency with which it uses its own share of the nation's water supply in lieu of division and transmission of water resources that could be more effectively used by other component of the society.

Water quality

Quality of water supply is important to livestock, farmsteads and other special irrigation uses. Imagine what life would be like if your local drinking water source was unsafe to drink, or there was not enough available for daily needs-if you had to rely only on bottled water brought from outside.

When water is drawn through a well, water moves through the aquifer from the area around it. The size of this area, called the *zone of contribution,* varies according to the rate of pumping and the direction of groundwater flow. It is very important to protect the zone of contribution from pollutant sources. Good planning and management are needed in order to maintain high water quality into the future.

Sources of water

Two major sources of water especially for agricultural use include surface and groundwater sources.

Surface water

The following are the major sources of surface water:

1. Rivers,
2. Streams and rivulets,
3. Surface wells,
4. Waterfalls and
5. Storm water with no definite paths.

For small supplies, water is pumped from the river or lake and piped to where it is required e.g. livestock keeping or small irrigation schemes. Pumping of water is subject to some fluctuations ranging from uninhibited flow during the rainy season to little or no flow when the rain have ceased. In order to prevent these fluctuations, structures such as dams are built to impound water during the time of uninhibited flow.

Groundwater water

Beneath our feet, in the sand, lie our groundwater/ drinking water. The water-saturated soil, known as the *aquifer*, extends in some places to a depth of approximately 400 feet (130 meters), then grows more shallow toward the edges of the land as seen in the cross-section.

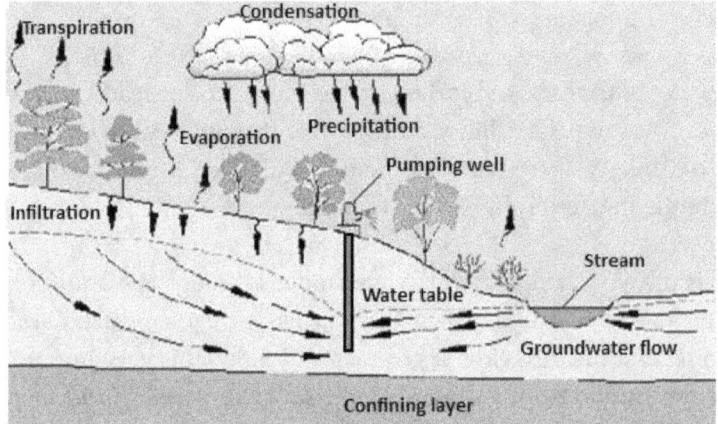

Figure 1-7: The hydrological cycle (*Louise Russell, 1999*)

All waters on the earth are continually in motion from the earth to the sky, driven by the water or *hydrologic cycle*. This cycle is made of four basic processes: precipitation, percolation/surface runoff, evaporation/transpiration, and condensation.

Recharge of groundwater supply by water spreading, recharge well, replenishment, irrigation and similar practices provide another means of water resources development.

Zones of aeration and saturation

Soils have zones of aeration and saturation. The *zone of aeration* is above the water table, where the openings in the soil are filled with air. In the *zone of saturation*, all the open spaces in the soil are filled with water. Plant roots must have aeration as well as moisture.

The concept of aeration and saturation may be simulated by a damp versus soaked kitchen sponge. The saturated sponge will accept no more water, and all air space had been replaced by water. Saturated zone and the unsaturated zone can be distinguished. Water contained in saturated zones is important for engineering works, geologic studies water supply development and petroleum engineering.

Utilization of groundwater: Its use in irrigation, industries, municipalities and rural homes continues to increase.

Terms associated with groundwater

1. *Aquifer:* This is a geologic formation having structures that permit appreciable amount of water to move through it under ordinary field conditions. They are in geologic formations like: limestone deposits, volcanic rocks, sandstones, crystalline and metamorphic rocks, and unconsolidated aluminum or rock materials derive from erosion of bordering mountains.

   Types of aquifers

   a. *Confined aquifer:* Confined aquifer is a body of water trapped between two confirming impervious layers. When such body of water is penetrated by a borehole or well, water rises into the well above the surround water table level.
   b. *Unconfined aquifer:* This is an aquifer in which water table serves as the upper zone of saturation.
   c. *Perched aquifer:* This is a special case of unconfined aquifer in which a body of ground water is separated from the main water body by an imperious layer.

2. *Aquiclude:* An impermeable formation which may contain water but is incapable of transmitting it in significant quantity. Clay is an example.
3. *Aquifuge:* This is an impermeable formation that neither contains water nor is capable of transmitting it. Granite is an example.
4. *Groundwater yield;* The quantity of ground water which can be withdrawn without impairing the aquifer as a water causing contamination or creating economic problem from a several increase pump lift.
5. *Ground water recharge:* Return of water into the soil to replace what has been abstracted in order to increase ground water yield. Water enters the formation from the ground surface or from the body of surface water after which it travels slowly to the surface by action of natural flow, evaporation or extraction by man.

1.3   Water well and borehole

Water well is a vertical hole excavated in the earth for bringing ground water to the surface. Shallow wells are either dug, bored, drived or jetted; however, deep wells are drilled by cable tools, hydraulic, rotary or reverse rotary methods.

Testing boreholes and well logs

It is a common practice to put down test borehole to determine depths of ground water, quality of water and physical characteristics and thickness of aquifer etc before drilling well in new areas. The diameter of such test borehole is between 8-10 inches. During drilling of test boreholes, a record or log is kept of the various formations and the depth of which they are encountered. Careful analysis further helps to obtain information about the stratum.

Types of water well

1. *Shallow well*: Typical examples of shallow wells include:
   a. Dug *wells*: These are the commonest and vary in depth from 10 to 40 feet depending on the position of the water table. They are hand drilled and lined with casing of wood, brick or concrete and equipped with hand operated fetcher for water lift. It can yield between 500 – 1500gpm.

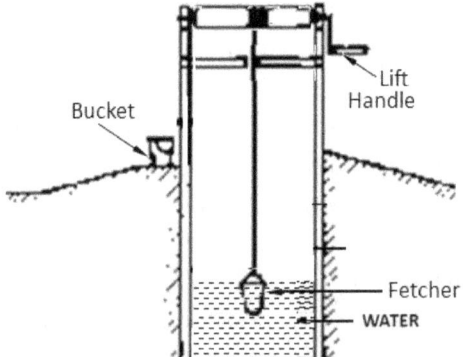

Figure 1-8: Bucket lift in closed well (Erik Nissen-Petersen)

   b. *Bored well*: Bored well exists where the water table is at shallow depths in an unconsolidated aquifer. They are usually constructed with hand operated hangers of varying diameters up to 72 mm (6") and depth up to 7200 mm (50ft) or power driven hangers, 432 mm (36") in diameter and up to 14400 mm (100ft) depth.
   c. *Driven wells*: This consists of a series of connected lengths of pipes driven by repeated impacts into the ground to below the water table. Water enters the

well through a driving point at the lower end of the well. A steel cone protects the screened range between cylindrical sections at the bottom. Diameter range between 1¼ "– 4" and yield between 20–50gpm.

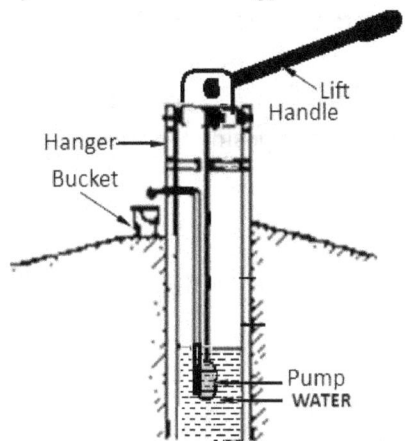

Figure 1-9: Pump driven well (Erik Nissen-Petersen)

d. *Jetted wells*: Cutting action of a downward stream jut at high velocity washed the earth away while casing is lowered into the deepening hole diameter is between 1½" – 3" and depths up to 50ft. The yield is small.

2. *Deep wells*

Most deep and high capacity wells are constructed by drilling. Three basic methods are employed in such construction as follows:

a. Cable tool method (Peroussion method)
b. Hydraulic rotary method and
c. Reverse rotary method

Each method is particularly suited for drilling in certain formations.

Well completion (development and test for yield)

Well completion provides for easy entrance of ground water into the well with minimum resistance. Perforated or screen casing used at the back provide filtration of water. Gravel packing is also often used. Following well completion a new deep well has to be developed to increase its specific capacity, prevent sanding and obtain maximum well life. Compressed air can be used to loosen the fine materials surrounding the discharge pipe. The yield and drawdown is tested for after development. A test is accomplished by measuring the static water level after which it is pumped at a maximum rate until the water level stabilized. The depth is noted.

The difference in depths is drawdown and the discharge drawdown ratio is an estimated specific capacity of the well.

1.4   Dam designs and development

Dams are usually made of earth, rock, concrete or masonry; therefore dams are classified according to material composing the structure as follows:

1. Earth dams
2. Concrete dams
3. Rock-fill dam

The choice of material depends upon:

1. The geology of the dam site
2. The cost of various alternatives

Foundation material below the dam must be water tight, or capable of being made water tight by such means as grouting.

Purposes for which dam may be developed

Dam projects are developed for the following purposes among many others:

1. *Domestic and municipal purposes*: Dams are constructed to serve such purposes as drinking, household use, gardening, park etc.
2. *Industrial uses:* Dams could serve water for production uses in factories, namely, breweries, distilleries, food preparations, bottling industries etc.
3. *Agricultural uses*: Dams are equally useful for irrigation purposes, dairying, inland fishery and industries.
4. *Power development*: Dams also serve as power source in hydro power generation and distribution.
5. *Flood control:* Dams could be used in control of excess water from flood, or swamp
6. *Wildlife*: Water is impounded in dams to provide water for wildlife – birds and animals.
7. *Recreation*: Dams serves water requirement in boating, swimming development parks and tourist attractions.
8. *Stream flow regulations*: Dam can be constructed across big streams or rivers to control flow

9. *Water conservation purposes*: Dams can be used for ground water recharge where water is impounded, and allow infiltration and percolation opportunities.

Feasibility studies of dam development

Thorough studies should be carried out to ascertain the success, usefulness, and the economic soundness of the dam project. The feasibility of engineering work, wisdom of investment, the overall benefit to the owners among other things should be investigated.

Salient items to investigate include:

1. *Site specific condition* with respect to:
   a. Soil stability
   b. Adequate impoundment area
   c. Suitable reservoir location
   d. Suitable construction materials etc.
2. *Hydrological data*

Hydrological data of the location include; flow records, flood record studies, sedimentation and water quality studies, ground water table in the vicinity of dam or reservoir, water rights, climatologic data (evaporation , temperatures, wind), construction costs, maintenance costs, economic benefits, social benefits, unquantifiable benefits such as scrutiny consideration, national pride etc. A report justifying the project or suggesting alternatives should be written following the investigations.

Selection of dam type

The following physical factors govern the selection of dam type.

1. *Topography*: Location of the spillway will be governed by local topography and material bearing on the final selection on the type of dam.
2. *Geology and foundation conditions*: This depends on the geological characteristic and thickness of the strata which are to carry the weight of the dam. This foundation limits to an extent the choice of dam.
3. *Construction materials*: Availability of construction materials, such as soils for embankment and sand at or near the site affects the type of dam.
4. *Spillway size and location*: Spillway size, type and the natural restriction in its location may be the controlling factor in the choice of the type of dam. The cost of

large spill way is a considerable portion of the total cost of development of water facility.

5. *Earthquake*: In areas prone to earthquake, earth fill dam comes to mind as first choice.

Gravity dam design considerations

The following considerations should be satisfied in order to design a stable gravity dam

1. Reservoir must be water tight at least to the level of the intended top water level of reservoir.
2. Foundation must be strong enough to resist all forces coming upon them.
3. Dam must be properly connected to water light foundation.
4. The dam and all its work must be durable
5. Provision must be made to pass all flood water safely past the dam.
6. Provision must be made to draw off water from the reservoir under control for supply purposes.

Forces acting on dam design

There are forces acting on dam structures. These forces are transmitted to the foundation and abutment of the dam which react against the dam with an equal and opposite force. A dam must be relatively impervious to water and capable of resisting the forces acting on it.

These forces include:

1. Gravity (weight of the dam)
2. Uplift
3. Earthquake forces
4. Hydrostatic pressure
5. Ice pressure (in temperate regions).

1. *Gravity (weight):* This is the product of the area of the dam and the specific weight of the material of construction. The line of action of this force passes through the centre of the area of the cross section.

$$W = \gamma_c A \dots \dots \dots \dots \dots \dots \dots \dots \dots \dots \dots \dots 1.4$$

Line of action of the weight of dam Figure 1-8

$$W_1 = 2/3(b - t) \ldots\ldots\ldots\ldots\ldots\ldots\ldots .1.5$$

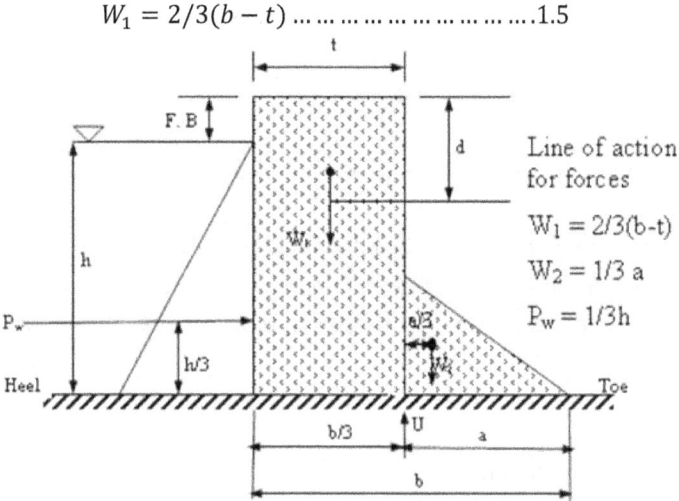

Line of action
for forces

$W_1 = 2/3(b-t)$

$W_2 = 1/3\ a$

$P_w = 1/3h$

Figure 1-10: Forces acting on a dam structure

2. *Hydrostatic force*: It may act on both the upstream and downstream faces of the dam. It has two components. Horizontal component of force on a vertical projection of the face of the dam.

$$P_w = H_h = 1/2\gamma_w h^2 \ldots\ldots\ldots\ldots\ldots\ldots\ldots .1.6$$

Where
γ_w = specific weight of water
h = depth of water.

The line of action of this force is acting at $1/3h$ above the base of the dam.

The vertical component of the force is the weight of water vertically above the face of the dam and passes through the centre of gravity of this volume of water.

$$H_v = \frac{1}{2\gamma h b} \ldots\ldots\ldots\ldots\ldots\ldots .1.7$$

3. *Uplift*: Pressure created by water hitting the surface of dam and its foundation and tends to lift up the weight of the dam off its foundation.

$$U = \frac{\gamma(h_1 - h_2)t}{2} \ldots\ldots\ldots\ldots\ldots\ldots . 1.8$$

Where

t = base thickness of the dam

h_1 and h_2 = water depths at the heel and toes of the dam respectively.

Design criteria of gravity dam

Gravity dams should be designed to satisfy the following conditions

1. *No overturning*: Reservoir is subjected to overturning effect due to external forces acting on it except the self weight which resists the over turning. The resisting and overturning moment taken about the toe must be equal to ensure safety against overturning, in which case

$$\sum \frac{M_t}{M_e} \geq 1.5 \dots \dots \dots \dots \dots \dots \dots \dots \dots .1.9$$

2. *No sliding*: Sliding is caused by the base shearing off or by lateral thrust ΣH acting on the foundation. Friction resistance offered by the foundation resist the sliding force.

For safety

$$\frac{\Sigma V}{\Sigma H} \geq 1.5 \dots \dots \dots \dots \dots \dots \dots 1.10$$

3. *No tension*: Concrete is weak in tension thus there is tensile crack when there is too much tension leading to uplift pressure.
4. *No crushing*: The toe of dam is subjected to excessive internal stress at full reservoir condition, and the heel in empty condition. To ensure safety against crushing.

$$T_{max} > F_{permissible} \dots \dots \dots \dots \dots \dots \dots 1.11$$

The $F_{permissible}$ can be found by dividing the crushing strength by a suitable factor of safety usually 3 in case of elastic design.

5. *No foundation failure*: This takes place on the plane of the contact surface, at a weaker stratum below the plane of contact.

Earth dams

Earth dams are dams built from natural earth materials including fill materials, ripraps, gravels etc. as well as cement or other embankment materials. Earth dams are distinguished by the placement of the different earth material at different part of the dam as follows:

1. *Simple homogeneous embankment* – Single exclusive impervious materials for slope protection and water barrier.
2. *Zones embankment* – A central impervious cone is flanked by zones of material considerably more impervious.
3. *Diaphragm embankment* – The bulls of the embankment is constructed of pervious materials and a thin diaphragm of impermeable materials to form barrier for water. The diaphragm may consist of earth, cement concrete or other materials.

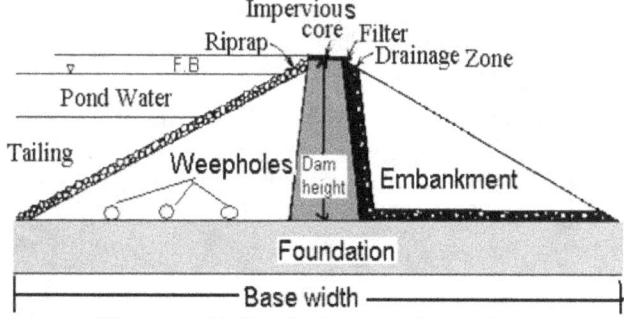

Figure 1-11: Earth dam configuration

Earth dam failure

Failures of many earth dams due to poor design have made it apparent that earth dams require as much engineering skills in conception and construction as any other type of dam.

Failure of dams owning to slipping of the embankment may be caused by one or multiple of the following factors:

a. Too steep a slope
b. Poor frictional resistance of the foundation and embankment soil e.g. claying and silty materials.
c. High poor pore pressure development due to inadequate drainage of the water use during compaction of the core material.
d. Steady seepage from upstream to downstream face.

e. Seepage due to sudden drawdown of the water level in the reservoir. For instance, the failure of Baguada dam in Kano in 1988 was caused by seepage through piping damage. A drawdown of the dam over a 3 year period promotes termite activities resulting in holes and small craters within the dam structure thus initiating piping when the dam was re-impounded.

Seepage losses in dams

1. Seepage could occur through strata or pockets or permeable materials. These may or may not be exposed during excavation (See Figure 1-12 (1)).
2. Seepage also could occur along roots and root cavities. If trees in the vicinity of the embankment are cut, the roots will die, leaving cavities through which water will flow (Figure 1-12 (2)).
3. Seepage along the plane between the original ground and the embankment fill material (Figure 1-12 (3)).
4. Seepage also occurs under the embankment and through a layer of permeable material (Figure 1-12 (4)).
5. Seepage equally occurs through the embankment (Figure 1-12 (5)).

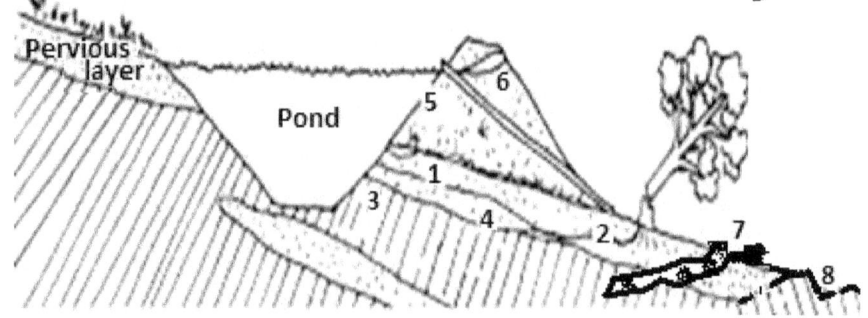

Figure 1-12: Situations in which seepage may occur from a water storage structure

6. Seepage along pipes passing through embankment (Figure 1-12 (6)).
7. Flow through muskrat burrows and cavities created by other burrowing animals (Figure 1-12 (7)).
8. Seepage over the entire basin at sites where the soil is permeable throughout the profile (Figure 1-12 (8)).

Precaution against seepage losses during construction

The following precautions could guide against seepage from water storage structures:

1. Make a thorough site investigation prior to construction. Either move to another site or give special treatment, such as installation of a compacted blanket of impervious material (See Figure 1-13 (1)).
2. Remove all roots from the embankment area prior to construction (Figure 1-13 (2)).
3. Remove all debris and sod and scarify soil surface before adding fill. Possibly construct core trench.
4. Block flow by construction of a core trench (Figure 1-13 (3)).
5. Build the embankment with proper top width and side slopes; remove all brush, roots, and debris from the borrow area so it will not be deposited in the embankment; place the less permeable material on the water side of the embankment and the more permeable material in the downstream part; place fill in thin layers and compact thoroughly.

Figure 1-13: Precautions against seepage *in* water storage structures

6. Install anti-seep collars and properly bed the pipe and compact earth around it.
7. Build embankments with proper top width and side slopes; manage the lagoon to minimize water-level fluctuation; keep the embankment clear of brush and debris.
8. Scarify the basin area to a depth of 8-10 inches, compact the loosened soil at optimum moisture content to form a dense layer, and on more permeable soils, install a blanket of compacted low-permeability earth.

Controls for seepage structures

1. *Cut off trenches* – should extend from the bedrock foundation up stream.

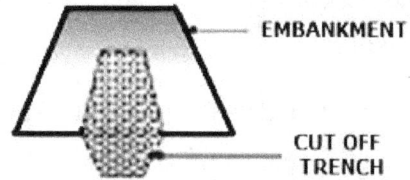

Figure 1-14: Control structures for earth dam

2. *Sheet piling cut off*: Used with partial cut off trench as an economic means of increasing the depth of cutoff and hence the path of seepage flow. Piling should be made of steel to increase strength.

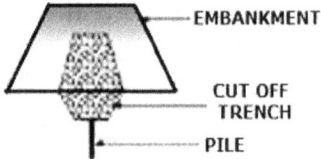

Figure 1-15: Control structures for earth dam

3. *Cement – bound – curtain cutoff*: This is a mixture of cement bound curtain in-place. Grout is pump through a hollow rotating drill-rod at the end of which is a mixing head.
4. *Grouting* – cement, asphalt clay and various chemicals (sodium silicate) is injected into the pervious foundation to increase stability and prevent seepage.

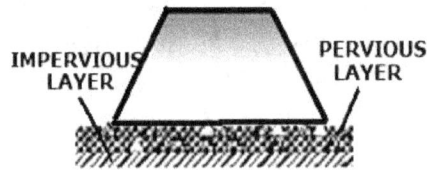

Figure 1-16: Control structures for earth dam

Design of earth dams

Earth dams should be designed to satisfy the following conditions

1. *Slope*: Embankment slope is decided based on experience and often ratio of Horizontal distance: Vertical rise

$$Slope = \frac{Horizontal\ distance}{Vertical\ rise} \quad \text{......................} 1.12$$

For upstream dams, the slope should be 3:1 and 2:1 or 2½:1 for downstream with flatter slope for higher dams.

2. *Height of dam*: Distance from foundation to the water surface when the spillway is discharging at designed capacity plus the free board allowance for wind, tide and frost action. It is determined by the volume of water impound without dangers, of hydrostatic pressure pushing too hard on the embankment plus available free board.
3. *Top width*: various expressions have been used to determine this factor. One of such is given by US bureau of reclamation proposal.

$$B = H_d + 10 \dots \dots \dots \dots \dots \dots \dots \dots \dots 1.13$$

Where

  B = top width

  H$_d$ = Height of dam in ft. Top width is designed to withstand shock and to keep the pressure line on upper surface of seepage within the dam when reservoir is full.

4. *Seepage*: - No earth dam is totally impervious; some amount of seepage is expected. If the rate of pressure drop resulting from seepage exceeds the resistance of the soil particle to movement the particle will move, resulting in piping – removal of fine particle from embankment.
5. *Pore pressure*: After dam construction, sizeable pore pressure will be present. If not removed, water in the pores of the dam caused a new pattern of pore pressure to develop.
6. *Slope stability*: Usually dam failure results from sliding of large mass of soil along a curved surface. The location of the centre of failure is got by assuming that the earth wall is divided into segments and each weight calculated. The movement of forces tending to rotate the soil mass about O is figured using formula.

$$M = \sum WX \dots \dots \dots \dots \dots \dots \dots \dots \dots .1.14$$

Where

  W = weight,

  X = moment arm of individual segment.

Tangential shear stress acting along the failure line can create a resisting moment M$_R$.

$$M_R = \Sigma S_s(\Delta L)_r \dots \dots \dots \dots \dots \dots \dots \dots 1.15$$

Where

  S$_s$ = Shear strength of the soil

  L = Length of a failure arc for a segment

  r = Radius of the failure arc.

7. *Slope protection*: Upstream slope should be protected against wave action by a riprap or concrete.

1.5 Spillways

Spillways are structures provided for storage and detention dams to release surplus or flood water which cannot be contained in the allotted storage space and at diversion dams to by-pass flow exceeding those which are turned into the diversion system. Water impounded by an embankment enters the spillway through a box, a weir in a wall, or a culvert-type entrance.

Types of spillways

Several forms of concrete or metal structures have been used as spillways; however a great number of designs have been employed. Few of the identified spillways design include:

1. *Free-over fall (straight drop) spillways:* In freefall spillways, water flow drops freely from the crest and it is found in thin arch or deck over flow dams.
2. *Two drop spillways:* Water impounded by an embankment enters the spillway through a box, a weir in a wall, or a culvert-type entrance and then flows to the downstream section.

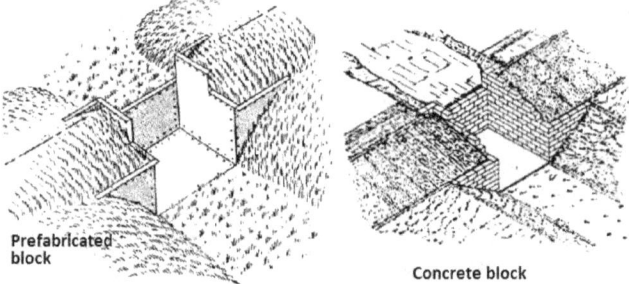

Figure 1-17: Two drop spillways (El-Swaify *et al.*, 1982)

3. *Drop inlet spillway* – water enters over a horizontally positioned tip and drop through a vertical or sloping shaft and then flows to the downstream river channel through a horizontal tunnel or the conduit.

Figure 1-18: Drop inlet spillway (El-Swaify *et al.*, 1982)

4. *Siphon spillways* – This is a close conduit system formed in the shape of an inverted tube positioned so that the inside of the band of the upper passageway is at normal reservoir storage level.

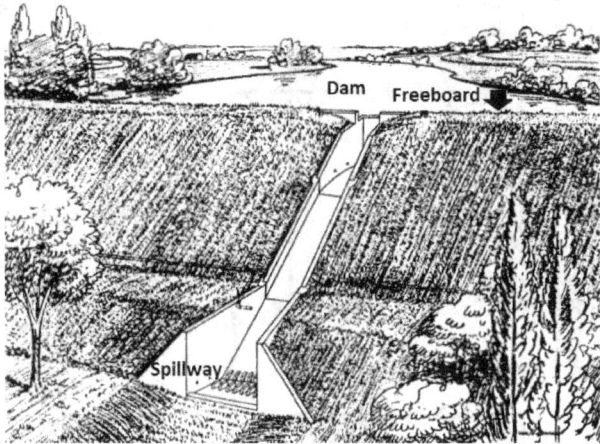

Figure 1-19: Chute spillway (El-Swaify *et al.*, 1982)

Spillway components

Spillway components include:

1. *Discharge channel* – Discharge channel dimensions are governed by hydraulic requirements, but selection of profile, cross-section, width, length etc are all influences by geologic and topographical characteristics off the site.
2. *Control device*: This is required to regulate outflow from reservoir. It could be sharp – crested or opee shaped and they discharge freely or partly freely or partly submerged.
3. *Terminal structures* –static flow fall from reservoir downstream level is converted to kinetic energy which manifests itself in form of high velocity which if impeded results in large pressures. The flow is returned to the dam through these structures.

Selection criteria for spillways

A general rule-of-thumb chart that could be used to determine the required spillway type is given in the chart in Figure 1-20. The chart shows most economical structure as related to discharge and controlled head providing adequate site conditions.

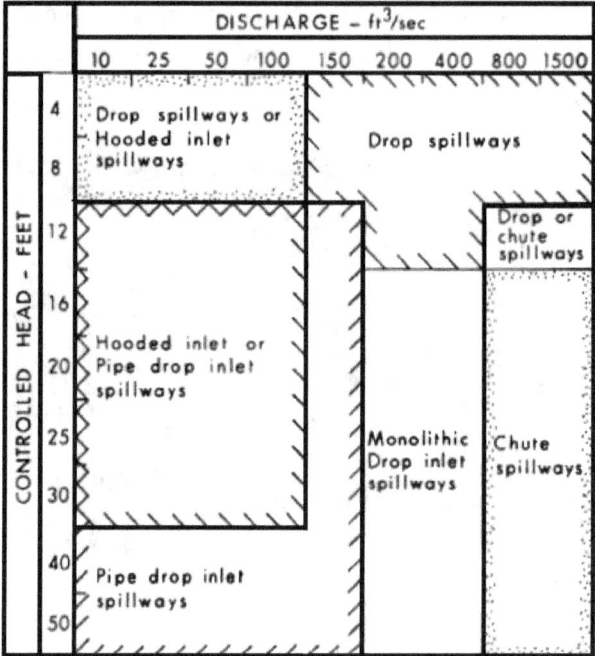

Figure 1-20: Spillway selection criteria (El-Swaify *et al.*, 1982)

1.6 Reservoir and its physical characteristics

A reservoir stores or conserves what could have been lost during precipitation at an earlier period. Primary function of reservoirs is to provide storage and its most important physical characteristics is storage capacity. On natural site, capacity is determined from topographic surveys, or computed by the use of Prismodal formula.

Characteristic features of a reservoir

1. *Normal pool level*: This is the maximum elevation to which the reservoir surface will rise during ordinary operation condition.
2. *Minimum pool level*: This refers to the lowest elevation to which the pool is to be drawn under normal conditions.
3. *Reservoir sedimentation*: Rate of sedimentation should be determined to determine whether the useful life of the proposed reservoir is sufficient to warrant its construction.
4. *Control*: Reservoir sedimentation cannot be prevented or avoided, but can only be retarded through the following ways:

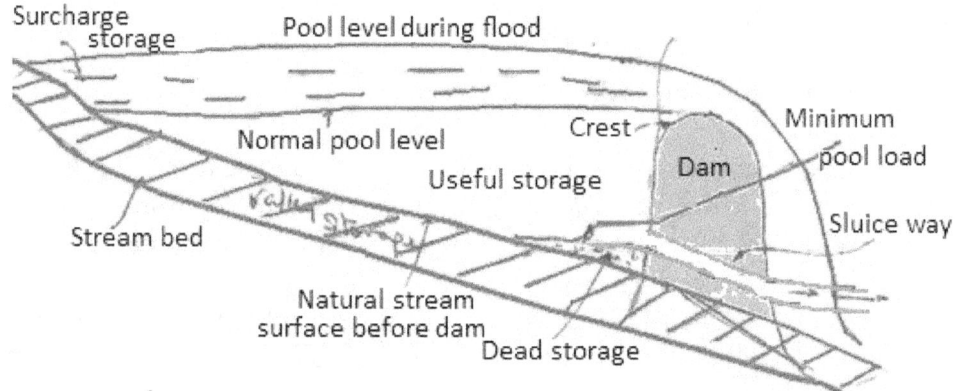

Figure: 1-21: Characteristic features of a reservoir

a. Select a site where sediment inflow is low
b. Use soil conservation methods within drainage basin e.g. terraces, contour ploughing etc.
c. Check dams in gullies, and prevent sediments from entering the stream.
d. Impact of rain drop is drop is controlled by vegetal cover
e. Construct sluice gates at various levels
f. Physical removal of sediment.

1.7 Pipe hydraulics

Water conveyance (pipe hydraulics)

Two types of conduits are used in conveying water thus: open channels and pressure conduit pipes

1. *Open channels*: Open channel is a conduit for flow which has a true surface. The free surface is essentially an interface between two fluids of different densities. Open channel flows are almost turbulent and unaffected by surface tension however in many cases of practical importance, such flows are density stratified i.e. flow varies in all direction. The importance of density stratification is that when stable density stratification exist, i.e. density increase with depth or lighter fluid overlies heavier flied the effectiveness of turbulence as a mixing mechanism is reduced. Example of open channel include canal etc.
2. *Canals*: These are open channel structures for water conveyance. They can be earth canals on lined canals and their shape varies from rectangular to trapezoidal. Free board is provided for in design to prevent wave action.

Hydraulics of open channel flow

Two types of flow are common in open channel flows:

1. *Laminar flow* – Individual fluid particles follow a straight line path: This is rarely encountered in hydraulic engineering practice. This flow is unaffected by the nature of the boundary surface thus there is not term giving indication of wall roughness in its analysis.
2. *Turbulent flow:* Individual particles do not follow straight line path but diverse and turbulent. This is most commonly observed type of flow.

Channel sections

Properties of channel sections are generally determined by the geometric shape of the channel and the depth of flow Figure 1-37.

1. *Depth of flow, d-* the vertical distance from the lowest point of the channel section to the water surface.

$$d = r\cos\theta \ldots\ldots\ldots\ldots\ldots\ldots\ldots.1.14$$

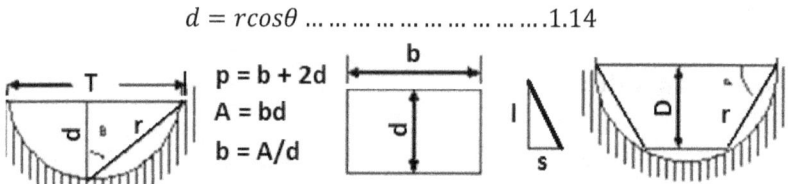

$$p = b + 2d$$
$$A = bd$$
$$b = A/d$$

Figure 1-22: Channel sections

2. *Channel slope, θ:* This is the intersection between the hydraulic radius and the hydraulic depth expressed in degrees.

$$\theta = \cos^{-1}\left[\frac{d}{r}\right] \ldots\ldots\ldots\ldots\ldots\ldots\ldots.1.15$$

3. *Hydraulic radius, r:* This is the ratio of the flow area to the wetted perimeter.

$$r = \frac{A}{p} \ldots\ldots\ldots\ldots\ldots\ldots\ldots\ldots.1.16$$

4. *Hydraulic depth d:* this is the ratio of the flow area to the top width

$$D = \frac{A}{T} \ldots\ldots\ldots\ldots\ldots\ldots\ldots\ldots.1.17$$

5. *Stage:* This is the elevation of water surface relative to a datum. Stage is equal to depth if the lowest part of channel is taken as datum.
6. *Top width, t:* The width of channel cross section at the water surface.
7. *Flow area, A-* the cross sectional area of the flow taken normally to the direction of flow.
8. *Wetted perimeter, p:* This is the length of the line of the interface between the fluid and the channel boundary.

Economic channel sections

It is of interest to deliver the maximum flow for a given cross sectional area of flow and slope when excavating a channel. To have the best economic channel for rectangular section, R must be equal to d/2. For trapezoidal section, the most efficient hydraulic section is when the top width is twice the length of a sloping side. i.e.

$$b = 2d\sqrt{(1 + S^2)} - 2Sd \dots \dots \dots \dots \dots \dots \dots \dots \dots 1.18$$

Thus the best economic channel for trapezoidal section, is when the hydraulic radius

$$R = \frac{d}{2} \dots \dots \dots \dots \dots \dots \dots \dots \dots \dots \dots .1.19$$

The following factors militate against the adoption of the best section channels.

1. High cost of suitable lining which serve to raise the permissible velocity
2. Bank stability
3. Sharp increase in excavation cost with depth.

Hydraulics of pressure conduits

Fluid flow in pressure conduit pipe is full; they are preferred for public water supply because of reduced opportunity of pollution.

According to Bernoulli's, the total energy head at any point in a pipeline is the sum of the elevation, pressure and velocity heads thus

$$Z_A + \frac{P_A}{\gamma} + \frac{V_A^2}{2g} = Z_c + \frac{P_c}{\gamma} + h_f \dots \dots \dots \dots \dots 1.20$$

Where

$$h_f = \lambda l \frac{V^2}{2gD} \quad \text{...... ... ...} \; 1.21$$

The total energy can be represented by the energy grade line. Neglecting minor losses the slope of the energy line or energy gradient is the same as the friction gradient h_f. The hydraulic grade line is determined by the sum of the pressure and elevation heads, measured relative to the centroid of the pipe section hence it lies below the energy line by an amount equal to the velocity head, $\frac{v^2}{2g}$ and represents the height to which liquid will rise in an open stand pipe connected to the pipeline at any point.. The knowledge of hydraulic gradient is of value since it indicates the pressure radiation to which the pipe line is subjected.

Flow measurement

Flow in large natural channels is usually measure with current meter. In small streams and, manmade channels flows are measured by weirs or venturi flumes e.g. Parshall flume. Flow is obtained by multiplication of velocity of water through the flume by the product of the height of water through the throat.

In two dimensional flows, measurement of the strength of the density stratification is given by Richardson Number, R_i

$$R_i = g \left[\frac{\partial p}{\partial y} \right] / \left[\frac{\rho \partial u}{\partial y} \right]^2 \quad \text{... ... ...} \; 1.22$$

Where
  g = acceleration due to gravity.
  ρ = fluid density
  y = vertical coordinate
  $\partial p / \partial y$ = gradient density in vertical direction
  $\partial u / \partial y$ = gradient velocity in vertical direction

When $\partial u / \partial y$ is small relatively to $\partial p / \partial y$, Ri is large and stratification is stable. When $\partial u / \partial y$ is large relative to $\partial p / \partial y$, R_i is small and as $R_i \rightarrow 0$, the flow system approaches a homogenous or neutral condition. A flow can be sub-critical, critical or supercritical depending on the magnitude of the inertia force to gravity. This is based on Froude's Number

$$F = \frac{U}{\sqrt{gl}} \quad \text{... ... ...} \; 1.23$$

Where

U = A characteristic velocity of flow

R= the characteristic length.

If F = 1, the flow is critical

F< 1 the flow is suborbital, gravitational force dominate

F> 1 the flow is supercritical, inertial force dominate

Pipe network

Piping system is used to distribute water to a large agricultural irrigation undertaken, a city or large industrial plants. In any pipe network, two conditions must be satisfied

1. The algebraic sum of the pressure around any closed loop must be zero
2. The flow entering a junction must equal the flow leaving it.

The following types of piping are possible;

1. *Compound pipes* - This piping consist of pipes of several sizes in series
2. *Looping pipes* – Looping pipe consists of two a more pipe which branch and come together again downstream (in parallel)
3. *Branching pipe* - two or more pipes which branch and do not come together again downstream.
4. *Series pipe*: Series pipe connection consists of pipes of different sizes joined together in one straight connection. Suppose that it is required to determine the discharge, Q in a series connection between two reservoirs where the difference in level of the two reservoirs is H given by:

$$H = h_{f1} + h_{f2} + h_{f3} \dots \dots \dots \dots \dots 1.24$$

$$Q = \frac{\pi\sqrt{2gH}}{4\left[\frac{\lambda_1 l_1}{D_1^5} + \frac{\lambda_2 l_2}{D_2^5} + \frac{\lambda_3 l_3}{D_3^5}\right]^{\frac{1}{2}}} \dots \dots \dots \dots \dots \dots 1.25$$

5. *Equivalent pipes* - A pipe is equivalent to another pipe or to piping system when for a given lost head the same flow produced in the equivalent pipe as was produced in the given system. The equivalent pipe length is given by:

$$L_e = D_e^5/\lambda_e \left[\frac{\lambda_1 l_1}{D_1^5} + \frac{\lambda_2 l_2}{D_2^5} + \frac{\lambda_3 l_3}{D_3^5}\right] \dots \dots \dots \dots \dots 1.26$$

Where

Le = Equivalent length

De = Equivalent diameter

λe = Equivalent frictional factor

6. *Parallel pipes*-There is a common pressure drop in parallel pipes, therefore

$$H = \lambda_1 l_1 \frac{V^2}{2gD_1} = \lambda_2 l_2 \frac{V^2}{2gD_2} = \lambda_3 l_3 \frac{V^2}{2gD_3} \dots\dots\dots\dots\dots\dots1.27$$

Flow in pipe (pipe discharge), Q

Total flow in pipes, Q is expressed as

$$Q = Q_1 + Q_2 + Q_3 \dots\dots\dots\dots\dots 1.28$$

Using the equivalent pipe concept this afford a convenient method of analysis therefore

$$L_e = \frac{D_e^5}{\lambda_e \left[\frac{\lambda_1 l_1}{D_1^5} + \frac{\lambda_2 l_2}{D_2^5} + \frac{\lambda_3 l_3}{D_3^5}\right]} \dots\dots\dots\dots\dots 1.29$$

$$L_e = [D_e^5/\lambda_e]^{\frac{1}{2}} - \left[[D_1^5/\lambda_1 l_1]^{1/2} + [D_2^5/\lambda_2 l_2]^{1/2} + [D_3^5 \lambda_3 l_3]^{1/2}\right] \dots\dots\dots 1.30$$

Example

Water is pumped from a reservoir A to a reservoir B through a piping system which consist of 610mm diameter pipe, 450m long branching into two pipes of diameters 305mm and 457mm, each 600m long, The pumping station is situated adjacent to reservoir A and the surface level of B is 60m above that of A. Determine the heads of the pump if water is to be transferred at the rate of 0.4cumes also determine the flow in each of the branched pipes. Take λ to be equal to 0.02 for all pipes.

Solution

The piping network is shown in Figure 1-23 below. The parallel pipes will be replaced by a single equivalent pipe 610mm diameter i.e.

$$L_e = [D_e^5/\lambda_e]^{\frac{1}{2}} - \left[[D_1^5/\lambda_1 l_1]^{1/2} + [D_2^5/\lambda_2 l_2]^{1/2}\right]$$

But $\lambda_e = \lambda_1 = \lambda_2 = 0.02$

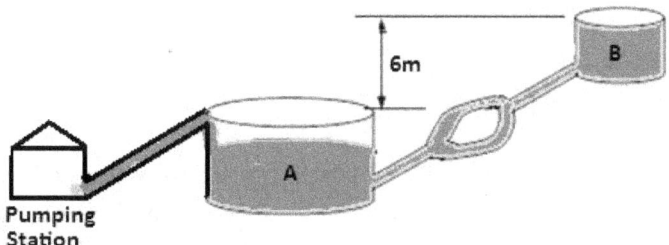

Figure 1-23: Pipe network

Thus,

$$[0.610_e^5/\lambda_e]^{\frac{1}{2}} = \left[[0.305_1^5/600]^{\frac{1}{2}} + [0.457_2^5/600]^{\frac{1}{2}}\right] \Rightarrow \lambda_e = 1367m$$

Total equivalent length is 1367 + 450 = 1817m.

Head loss $h_f = \lambda lV^2/2gD$ $\qquad = \lambda l_e/2gD \; [4Q/\pi D_e^2 = 5.69m$

$$h_f = \lambda l \frac{V^2}{2Dg} = \lambda l_c/2gD[\frac{4Q}{\pi D_e^5}] = 5.69m$$

Head loss in each of the branched pipes is the same

$$\therefore \quad \frac{Q_2^2}{D_2^5} = \frac{Q_3^2}{D_3^5}$$

$\Rightarrow Q_3 = 2.76 \; Q_2$ But $Q_2 + Q_3 = 0.4$ cumes(m³/s)

$\therefore Q_2 = 0.104$ cumes (m³/s), $\qquad Q_3 = 0.293$ cumes (m³/s)

Forces acting on a pipe

Pipes must be designed to withstand internal and external stresses, changes in momentum of flowing water, external loads temperature changes apart from satisfying hydraulic requirements. Internal pressures are caused by static pressures and water hammer. Internal pressure causes circumstantial tension in the pipe wall given by the formula.

$$\sigma = \frac{Pr}{t} \dots\dots\dots\dots\dots\dots.1.31$$

Where

σ = tensile stress,

P = pressure (static + water hammer)

r = Internal radius of pipe

t = thickness of the pipe wall

Water hammer – When a fluid flowing in pipeline is abruptly stopped by the closing of a valve, dynamic energy is converted into elastic energy and a series of positive and negative pressure-waves travel back and forth in the pipe. It is therefore important to ensure that pipeline is able to withstand the maximum or minimum pressure resulting from the value adjustment and other possible outlets. Also a practical solution is to provide automatic pressure release values and diversion conduit at the turbine installation.

Conveyance of irrigation and drainage water

Irrigated lands are often located at distances from the source of their water supply. Main conveyance canals of many irrigation projects vary from a few meters to several meters or more from point of use. Water obtained from streams and surface reservoir is conveyed farther than water obtained from underground reservoirs. Water delivery to distance reservoir is done by pumping.

1.8 Pumps

To get water delivered to the point of need, it has to be pumped. Water is pumped from the source and delivered through pipes to sprinklers. Costs are greatly reduced by obtaining the necessary energy for pumping from fuel rather than from human effort or animals. The horse power delivered by a motor or by an engine to the shaft of the pump is called *brake horse power*. The ratio of useful water horse power, WHP delivered by pumping (output) to brake horse power BHP, (input) from engine is known as *pump efficiency* (E.P).

$$E_p = \frac{WHP}{BHP} \ldots\ldots\ldots\ldots\ldots\ldots\ldots.1.32$$

Pump drivers

Pumps may be hand or power operated, designed to lift only or to lift and discharge against pressure and to lift from either shallow or deep wells. Pumps are driven by power are either electric motor, fuel engine, solar or wind mill.

Hand priming pumps: The simplest hand pump, often referred to as a pitcher pump, is satisfactory for use on wells or cisterns in which the water never needs to be lifted more than about 6m. A cross section of a pitcher pump is shown in Figure 1-24.

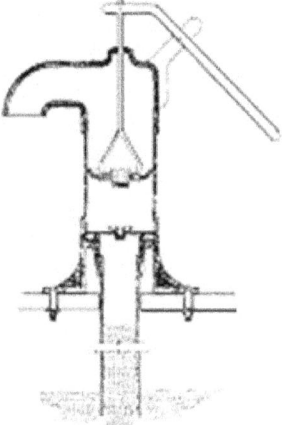

Figure: 1-24: Shallow well pitcher pump

Selection/classification of pumps

There are a number of pumps on the market from which to select for a particular application. They all have characteristics which influence their suitability for a specific water supply as well as the volume and pressure required.

The criterion for the selection/classification of type and construction of a pump is very important to meet the process specification and proper application. Knowledge of the variety of pumps in the market needed a review. There are two general types of pump in use in today's industry based on the principles of its operations: positive displacement and centrifugal (dynamic) pumps.

1. *Positive displacement* (PD) *pumps*

Positive displacement (PD) pumps work by allowing fluid to flow into some enclosed cavity from a low-pressure source, trapping the fluid, and then forcing it out into a high-pressure receiver by decreasing the volume of the cavity. Some examples of PD pumps are: fuel and oil pump in most automobiles, the pumps on most hydraulic systems, and the heart of most animals.

Some general types of the positive displacement pumps are as described below:

a. *Reciprocating pump*

Reciprocating pumps create and displace a volume of liquid, their "displacement volumes", by action of a reciprocating element. Liquid discharge pressure is limited only by strength of structural parts. A pressure relief valve and a discharge check valve are normally required for reciprocating pumps.

Reciprocating pumps are available for both shallow wells and deep-wells. They are capable of delivering water at quite high pressures. The shallow-well type is usually reasonable in cost, but the deep-well type tends to be expensive and it must be installed over the top of the well.

Reciprocating pumps can be further classified into three types as follows,

i. *Piston pumps:* Fluid is sent out by piston when the valve opens and water enters the upper cylinder. When the piston goes up the valve on the piston closes and water goes out through the channel.

ii. *Packed plunger pumps:* Water from deep wells is lifted with a similar plunger type pump in which the cylinder, including the plunger and valves, is supported on the discharge pipe deep enough in the well to be submerged in water at all times. The pump handle is connected to the plunger by means of a long rod. This type of pump is self-priming due to the cylinder being submerged in the water. Figure 1-25 illustrates a deep-well pump.

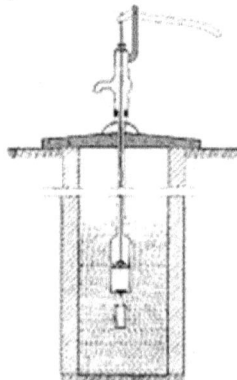

Figure: 1-25: Deep well plunger pump

iii. *Diaphragm pumps:* Diaphragm pumps have a piston and cylinder that are replaced with a diaphragm. As there are no sliding parts to wear, these pumps are suitable for pumping muddy water or high moisture slurries such as the waste from a biogas generator.

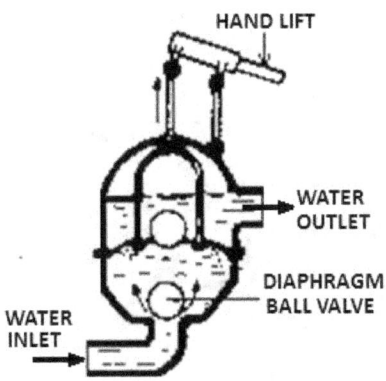

Figure 1-26: Diaphragm pump

b. *Rotary valve pumps*

Rotary pumps function with close clearances such that a fixed volume of liquid is displaced with each revolution of the internal element. This type of pump is popular for small quantities of water pumping. As the gears turn, they push small quantities of fluid. Rotary pumps include:

  i. Gear pump
 ii. Lobe pump
iii. Vane pump
 iv. Screw pump

All these pumps mentioned above have the similar working principles: pumping the liquid with the help of rotating elements. The difference lies on the rotating elements; they could be gear, lobe, vane, or screw.

2. *Centrifugal pumps*

Centrifugal pumps are dynamic pumps designed to raise the pressure of liquid by giving it a high kinetic energy and then converts it into pressure energy before the fluid exits the pump. It normally consists of an impeller (a wheel with blades), and some form of housing with a central inlet and a peripheral outlet (Figure 1-27). The impeller is mounted on a rotating shaft and enclosed in a stationary casing. Casings are generally of two types: volute and circular. The impeller design and the shape of the casing determine how liquid is accelerated though the pump.

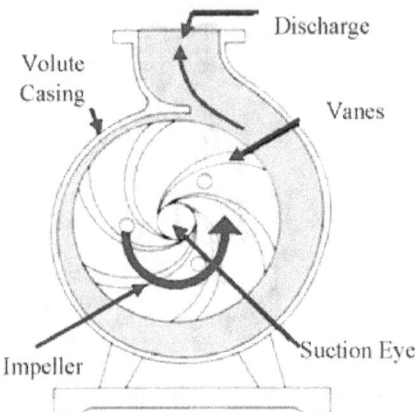

Figure 1-27: Liquid flow path inside a centrifugal pump

Centrifugal pumps are simple (only one moving part), durable, and relatively inexpensive for a given capacity. However, they are suitable only for low lifts of 3 to 4m and are prone to losing their prime unless the suction pipe is equipped with a good foot valve (check valve). Neither will they discharge against a very high head (pressure).

Horizontal centrifugal pump is eminently suitable for all water duties except those required in handling very large volumes of water against low head and pumping from the boreholes and wells.

Examples of centrifugal pump include:

a. *Turbo-mechanical pumps*: They are made of the casing, the impeller and the eye. Water enters from the bottom and goes out diagonally. They can either be:

  i. Mixed flow
  ii. Propeller type
  iii. Single or multiple stage and include such pumps as:

b. *Deep-well turbine pumps* are multi-stage centrifugal type and are driven either by a long vertical shaft from a drive head at the top of the well or by a submersible motor below the pump in the well.
c. *The submersible pump*, on the other hand, is available in a range of sizes and is an efficient, trouble-free design for medium-sized installations. Obviously it is a major operation to remove the pump from the well if something goes wrong. It should be noted that the motor is installed below the pump so that if the water level is reduced to the pump level, the motor will still be submerged in water which is essential for cooling.

d. *The impeller pump* may be an open type with a relatively large clearance between it and the casing or it may be a closed type with very close clearances. The open type will tolerate sand or silt in the water much better than the closed-impeller type. (Figure 1-28).

Figure 1-28: Impeller water pump

3. *Axial flow pumps*

Axial flow pumps are an established technology whose discharge of water is axial i.e. pumping is in a direction parallel to the pump shaft, rather than perpendicular as in centrifugal pump models (Figure 1).

Figure 1-29: Impeller water pump

The AFP is also widely referred to as a "propeller pump" because the impeller works much like a boat propeller (Figure 1-30). The water discharge capacity and fuel efficiency of an AFP are typically two to three times higher than a centrifugal pump at lifts below 3 m (IRRI 1983; Kathirvel et al. 2000 and Santos Valle et al. unpublished data).

Figure 1-30: Impeller showing the vanes

The most salient feature of the AFP is the long, solid pump shaft, which is actually a hollow pipe through which the rotating shaft passes. These shafts can be manufactured to various lengths and widths as needed - some pumps can be up to 7.5 m long. However, longer pumps are not necessarily designed to increase depth or water lift potential. Instead, longer AFPs have been developed so that the engine powering the pump can be safely set up on canal, river, or pond banks, without risk of losing the equipment to the water.

Pump usage

Centrifugal pumps are used in more industrial applications than any other kind of pump. This is primarily because these pumps offer low initial and upkeep costs. Traditionally these pumps have been limited to low-pressure-head applications, but modern pump designs have overcome this problem unless very high pressures are required. The single-stage, horizontal, overhung, centrifugal pump is by far the most commonly type used in the chemical process industry.

Basically, pump selection is made on the flow rate and head requirement and with other process considerations, such as material of the construction pumps for the corrosive chemical service or for the fluid with presence solids in the stream.

Choosing a pump

Five main factors must be considered when selecting a pump:

1. The total water required per day;
2. The maximum rate of flow desired;
3. The maximum flow from the water source;
4. The vertical distance the water must be lifted to the pump;

5. The total head against which the pump must operate.

Suitable system pump must be selected in line with maximum total head against which it works

$$H_t = H_u + H_m + H_s \,...............\, 1.33$$

Where

 H_t = Total design head against which pump is working (m)

 H_n = Maximum head required at main to operate sprinklers on laterals at required average pressure including riser height (m)

 H_m = Maximum friction loss in the main and suction line (m)

 H_j = Elevation difference between pump and junction of lateral and main (m)

 H_s = Elevation difference between pump and source of water after draw down (m).

Hydro-pneumatic systems

Hydro-pneumatic systems consist of an enclosed tank combined with an automatic pressure switch which turns the pump motor on when tank pressure drops to a preset level. As the tank is approximately half full of air, several liters of water can be pumped into the tank before the air is compressed and the stock cut-off pressure is reached.

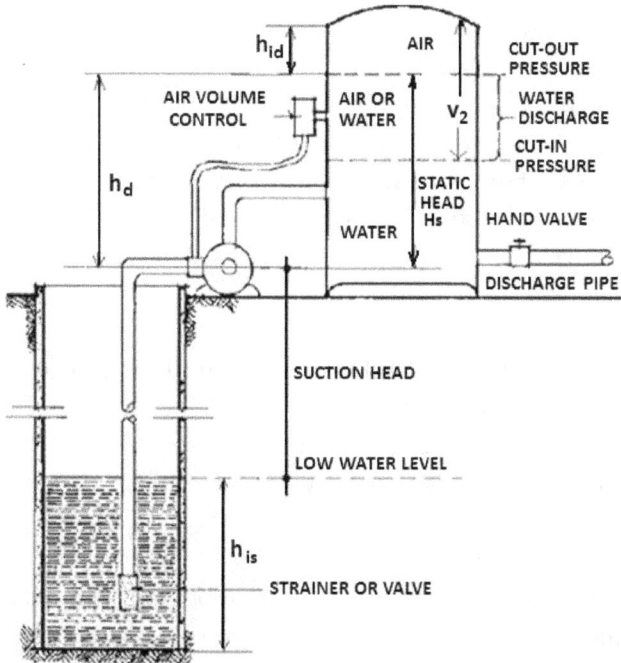

Figure 1-31: Hydro-pneumatic water system

The amount of water pumped into the tank can then be used as required before the pump needs to operate again.

There are several advantages to the hydro-pneumatic system:

- The tank can be located in any convenient place
- Optimum discharge pressure is available at all times
- The system is completely automatic
- The tank may be relatively small

Common definitions in pump calculation

Datum elevation – It is used as reference of the horizontal plane for which all the elevations and head are measured. The pumps standards normally specify the datum position relative to a pump part, e.g. centrifugal horizontal pump datum position is at the impeller shaft centerline.

Friction head-The head required to overcome the resistance to flow in the pipe and fittings. It is dependent upon the size, condition and type of pipe, number and type of pipe fittings, flow rate, and nature of the liquid.

Friction loss - Refers to reductions in flow due to turbulence as water passes through hoses, pipes, fittings and elbows.

Pressure head - Pressure head must be considered when a pumping system either begins or terminates in a tank which is under some pressure other than atmospheric. The pressure in such a tank must first be converted to feet of liquid. Denoted as h_p, pressure head refers to absolute pressure on the surface of the liquid reservoir supplying the pump suction, converted to feet of head. If the system is open, h_p equals atmospheric pressure head.

Dynamic discharge head- The static discharge head plus the friction in the discharge line also referred to as *total discharge head*.

Dynamic suction head - The static suction lift plus the friction in the suction line also referred to as *total suction head*.

Static suction head - Head resulting from elevation of the liquid relative to the pump center line (datum). If the liquid level is above pump centerline (datum), h_s is positive.

If the liquid level is below pump centerline (datum), h_s is negative. Negative h_s condition is commonly denoted as a "suction lift" condition

Static discharge head - It is the vertical distance in feet between the pump centerline and the point of free discharge or the surface of the liquid in the discharge tank.

Vapour pressure head - Vapour pressure is the absolute pressure at which a liquid and its vapour co-exist in equilibrium at a given temperature. The vapour pressure of liquid can be obtained from vapour pressure tables. When the vapour pressure is converted to head, it is referred to as vapour pressure head, h_{vp}. The value of h_{vp} of a liquid increases with the rising temperature and in effect, opposes the pressure on the liquid surface, the positive force that tends to cause liquid flow into the pump suction i.e. it reduces the suction pressure head. (Vapour pressure can be said as the external pressure require to prevent fluid from evaporate become vapour).

Velocity head - Refers to the energy of a liquid as a result of its motion at some velocity 'v'. It is the equivalent head in feet through which the water would have to fall to acquire the same velocity, or in other words, the head necessary to accelerate the water. The velocity head is usually insignificant and can be ignored in most high head systems. However, it can be a large factor and must be considered in low head systems.

Total head - Pressure required in feet (meter) of head that the pump must produce. The head at the discharge pump flange minus the head at suction flange.

Further reading

Amy Vickers, 2001. Handbook of water use and conservation" *water plow press,*

Beasly R. P., Gregory J. M., and McCarty T. R., (1984).Erosion and Sediment Pollution Control, 2d ed., (Ames: Iowa State University Press,), 242-43.

EI-Swaify S. A., Dangler E. W., and C. L. Armstrong, 1982. Soil Erosion By Water In The Tropics

FAO watershed management field manual gully erosion FAO conservation 13/2.

Fellenius W., 1979. Calculation of the Stability of Earth Dams, Trans. 2nd Congress on large Dams, Washington,.

John Lays, 2003. Wind erosion. Pub. Centre for Natural Resources NSW Department of Infrastructure, Planning and Natural Resources Parramatta. September 2003 ISBN 0 7347 5399 3 CNR 2003.069

Karl Kolmetz, 2012. Pump Selection and Sizing (engineering design guideline) KLM Technology Group, #03-12 Block Aronia, Jalan Sri Perkasa 2 Taman Tampoi Utama 81200 Johor Bahru.

Multiquip, 2012. Soil Compaction: a basic handbook. mq_soil_handbook.pdf. date assessed: 04/04/2012

Ozara N.A and Madubuike C.N. (1994). Biotechnical measures for erosion control on agricultural fields. Proceedings of the 1st S.E. Na/da Pre-season workshop organized under NALDA NAERLS memorandum of understanding.

SCS (Soil Conservation Service, USDA), *1975a.* Engineering field manual. Washington, DC: USDA.

Texas Agricultural Extension Service (2003). Agricultural water conservation practices. *Www.twdb.state.tx.us/assistance/conservation/agricons.htm*

Texas water development board. (2003) Agricultural Water Conservation Practices *www.twdb.state.tx.us/assistance/conservation/agricons.htm*

Van Morrill and Gabrielle Belfit (Eds.) (1999) CAPE COD-A Community Connected By Water. Part One, Water Resource Education Information brochure, November 1999

Weidelt, H. j., 1975 (compiler). Manual of reforestation and erosion control for the Philippines. Eschborn , West Germany: German Agency for Tech. Coop. Ltd.

Welter H. Wischmier and Dwight D. Smith (1976) predicting rainfall erosion losses. A guide to conservation planning US department of agriculture, Agriculture handwork No 537.

CHAPTER 2

Soil and Water Conservation Practices

2.0　Soil and water conservation engineering

Soil and water conservation engineering deals with the application of engineering principles to providing solutions to soil and water management problems. The conservation of these vital resources implies utilization without waste so as to make possible a high level of production, which can be, continued indefinitely.

Conservation is an important part of meeting agricultural soil and water demands in the years ahead. For instance, conservation should include design of field for efficient water use through good land preparation, landform practice, creating furrow dikes to conserve rainwater, and retaining soil moisture through conservation tillage. The loss of these resources and practices relevant to their conservations are discussed in the following sections.

2.1　Soil losses

Soil loss is the gradual or rapid washing or wearing away of the agricultural soil thereby making the field unsuitable for agricultural purposes. In more critical situation, the soil is degraded beyond the top soil, cutting through the subsoil. The importance of soil losses is indicated by the effect of top soil depth on crop yield and on some soils, this crop yield decreases and can largely be overcome by high fertilization.

Terms associated with soil loss

1. *Erosivity*: This is the power of rainfall to erode (cause erosion) or the ability of rain to form erosion. This depends on the kinetic energy of the rain drops. The amount of erosion caused by rainfall depends on three major factors namely:

    a. Nature of rain
    b. Kind of soil
    c. The way the land is managed

2. *Erodibility:* This is the extent to which the soil is vulnerable to erosion or the ability of the soil to be detached by the power of erosion.

Universal soil loss equation (USLE)

This equation is useful to determine the adequacy of conservation measure in farm planning and to predict non-point sediment losses in pollution control programme. Soil loss equations were developed to enable conservation experts to project limited erosion data to the many localities and conditions that have not been directly represented in the research.

The soil loss equation is

$$A = f(R, K, L, S, C, P) ton/acre/year \dots \dots \dots \dots \dots \dots \dots \dots \dots .2.1$$

$$A = 2.24(R, K, L, S, C, P) mg/ha \dots \dots \dots \dots \dots \dots \dots \dots \dots .2.2$$

Where
  A = Computed soil loss per unit area tons/acre/year
  R = Rainfall and runoff factor = number of rainfall erosion index units plus a factor for runoff from snowmelt or applied water where such run off is significant.
  K = Soil erodibility factor = soil loss rate per erosion index unit for a specified soil as measured on a unit plot, which is defined as a 72.6ft (22m) length of uniform 9% slope. If K is in mg/ha change constant K from 2.24 to 1 i.e. 1ton per acre = 2.24mg/ha.
  L = Slope length factor = ratio of soil loss from the field slope length to that from a (22m) length under identical condition.
  S = Slope steepness factor = ratio of soil loss from the field slope gradient to that from a 9% slope under otherwise identical condition.
  C = Crop management factor = ratio of soil loss from an area with specific cover and management to that from an identical area in tilled continuous fallow.
  P = Conservation practice factor = ratio of soil loss with a support practice like contouring to that with straight – row farming up and down the slope.

2.2   Soil engineering problems

Soil has exceedingly fragile zones where all or man's food, fiber, energy and industrial crops are cultivated and on which all of man's livestock are reared (Figure 2-1). Hence these activities results into soil degradation within these zones. Some of these activities constitute engineering problems which must be solved if soil is to remain relevant to sustainable agriculture.

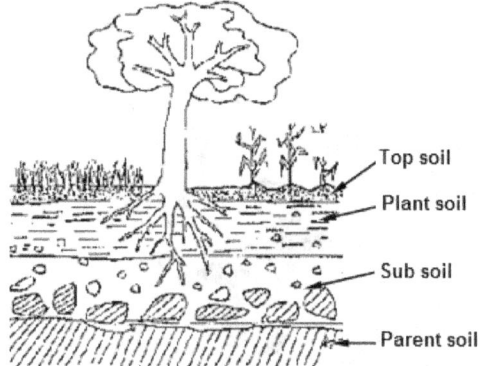

Figure 2-1: Different soil zones

The engineering problems involving soil can be divided into the following categories.

1. Human activities
2. Soil erosion activities
3. Soil compaction
4. Soil moisture conservation
5. Excess water (flooding) control

2.2.1 Human activities on agricultural soil

Human activities on agricultural soils have led to a substantial loss of agricultural top soil. Such activities as continuous farming, overgrazing, unfavourable agricultural practices, deforestation and lumbering among others over a long period of time are typical agents of soil degradation. Adoption of conservation farming systems rather than conventional system will help solve these problems to a large extent.

2.2.2 Soil erosion

Erosion is a process by which the soil particles are washed off the surface of the earth. Erosion is the gradual wearing way of soil particles from earth surface. Soil losses vary considerably with different types of erosion.

Two types of soil erosion exist:

a. *Geological erosion*: This includes soil forming as well as soil eroding processes, which maintain soil in a favorable balance suitable for the growth of most plants.

Example, weathering of rock is an erosion process; deposition of weathered material is a soil forming process

Figure 2-2: Soil erosion and degradation

b. *Accelerated erosion*: this includes the deterioration and loss of soil as a result of man's activities. It is normally associated with changes in natural cover soil conditions and it is caused primarily by water and wind. Forces involved in accelerated erosion are:

1. Attacking forces which removes and transport soil particles
2. Resisting forces which retarded erosion.

Types of erosion

Soil erosion occurs through two major agents; water and wind

Water erosion

This is the removal of particles from the soil surface by running water either as a result of massive water flow, excessive rainfall or flooding including run off from melted snow from icebergs. Water erosion is subdivided into raindrop, sheet, rill, gully and stream channels.

1. *Raindrops erosion*: Soil splash resulting from impact of water drops directly on soil particles or on thin water surfaces. On bean soil, it is estimated that as much as 100 tons of soil per acre are splashed into the air by heavy rains.

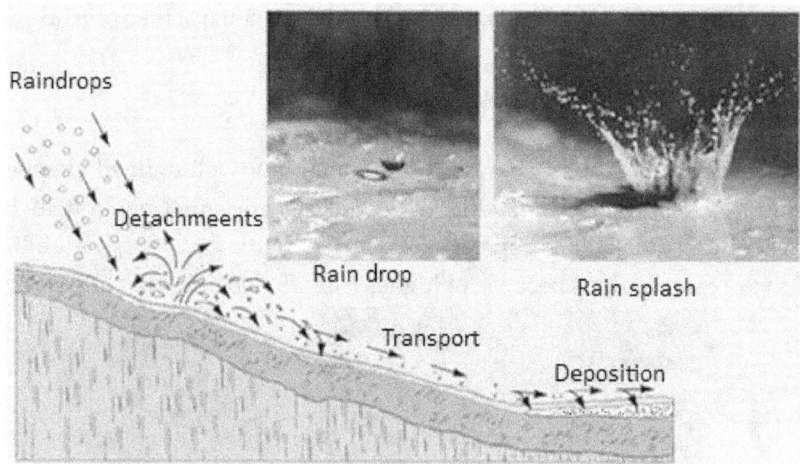

Figure 2-3: Signs of sheet erosion on unprotected farm

On level land, rain drops splash is not serious but on sloping fields, considerable amount of soil is splashed down hill factors affecting the direction and distance of soil splash are: slope, wind, surface condition and impediment to splash such as vegetative cover and mulches.

2. *Sheet erosion*: The idealized concept of sheet erosion is the uniform removal of soil thin layers from land resulting from overland slopes. Latest fundamental studies of the mechanism indicate that this form of erosion rarely occurs.

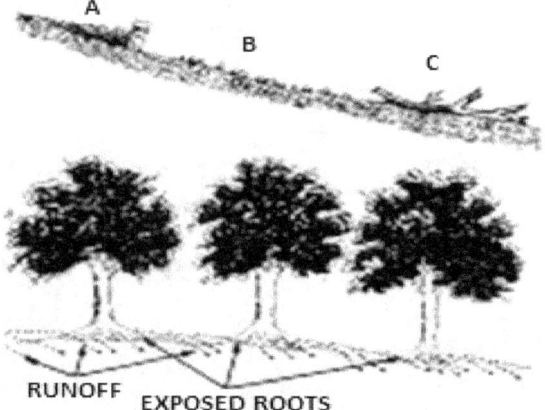

Figure 2-4: Sheet erosion in forest

Sheet erosion does leave visible marks Figure 2-4; such as soil collecting behind obstructions on slope A; stones left behind by runoff, B, or mouldy mounds of soil and other debris trapped under branches C,. The effect of sheet erosion is more apparent in forest areas without grass cover. The rainfall-runoff relationship from a given rainstorm is a function of:

i. Rainfall intensity distribution and sequence, during a particular rainstorm event;
ii. Soil infiltration rates; and
iii. The soil surface storage capacity.

3. *Rill erosion*: Removal of soil particles from small but well-defined channels when there is a concentration of overland flow. Rills are small enough to be easily removed by normal tillage operations. Detachability and transportability in rill erosion are more serious because of higher run off velocities that are involved.

Figure 2-5: Rill erosion

4. *Gully erosion:* Gully erosion produces channels larger than rills. These channels cannot be obliterated (removed) by tillage thus it is an advanced stage of rill erosion. The rate of gully erosion depends primarily on the run off producing characteristics of the watershed, the drainage area, soil characteristics, the alignment, size and shapes of the gully and the slope of the channel.

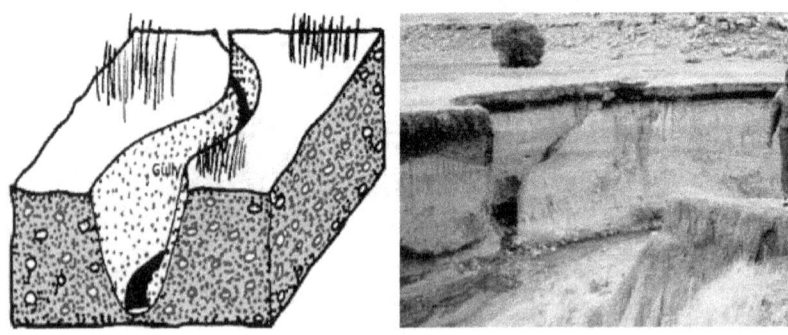

Figure 2-6: Schematic illustration of gully erosion

Gully formation

Four stages of gully development are generally recognized as follows.

Stage 1: Channel erosion by downward scour of the topsoil.

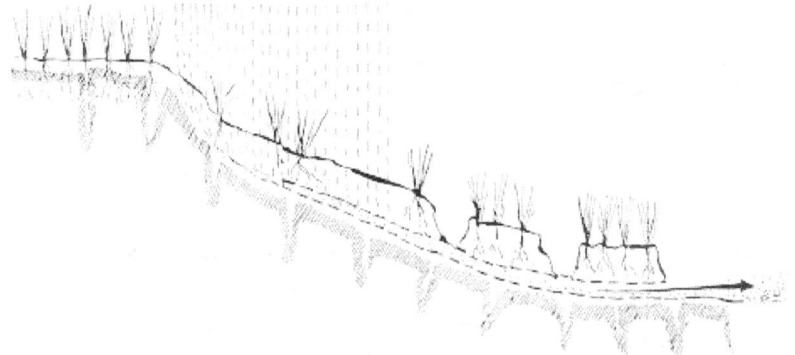

Figure 2-7: Gully tunnel at the lower edge of the watershed

Stage 2: Upstream movement of the gully head and the enlargement of gully width and depth as water advanced. The C-horizon is cut by the gully and weak parent material is rapidly removed

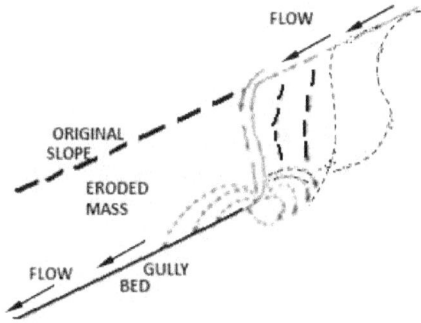

Figure 2-8: Advancement of gully to the upper edge of the watershed

Stage 3: *Healing stage;* with vegetation beginning to grow in the channel.

Stage 4: *Stabilization of the gully:* The channel reaches a stable gradient, gully walls reaches a stable slope and vegetation begins to grow in sufficient abundance to anchor the soil and permits development of new top soil.

Evaluation and prediction of gully development is difficult because the factors are not well defined and field records of gulling are inadequate.

Gully classification

Gully classification can be based on two major factors as follows;

a. *Gully size and drainage area:* Gully size and drainage area is shown in Table 2-1 below

Table 2-1: Gully configurations

Gully relative size	Gully depth (m)	Drainage area (Ha)
Small gully	1 or less	2 or less
Medium fully	1 to 5	2 to 20
Large	5 and above	20 and above.

b. *Gully cross section*: this may be U or V or trapezoidal shape depending on soil and climatic condition, age of the gully and time of erosion.

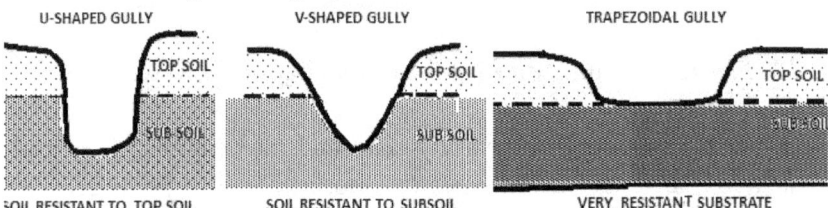

Figure 2-9: Shape of gully cross-sections

Factors affecting gully formation

Factor affecting gully formation can be categorized into two groups, man-made factors, and physical factors. Man-made factors include:

c. Improper land use
d. Forest and grass fires (bush burning)
e. Overgrazing
f. Mining
g. Road construction
h. Livestock and vehicle trails
i. Destructive logging (deforestation without replacement)

Physical factors affecting gully erosion

Physical factors also affect soil loss. The major variables include climatic factor such as precipitation, temperature, wind, humidity, solar radiation.

1. *Soil*: soil physical properties affect infiltration capacity and the extent to which it can be detached and transported. Such properties include texture, soil structure, moisture content, density or compaction and chemical and biological characteristics of the soil.
2. *Vegetation*: Major effects of vegetation in soil erosion are:

a. Interception of rainfall by absorbing energy due to raindrop impact.
b. Retardation of erosion by decreased surface velocity
c. Physical restrain of soil improvement
d. Increase biological activities in the soil
e. Increase soil moisture storage capacity

3. *Topography*: feature that influence erosion are
   a. Degree of slope
   b. Length of slope
   c. Size and shape of watershed.

a. Wind erosion

In the arid and semi arid regions of the world, large areas are affected by wind erosion. It removes soil and damage crops, fences, buildings and highways. Many humid regions are also damaged by wind erosion. The areas prone to damage are sandy soils, along streams lakes and coastal plains.

Soil movement

Three movements occur in wind erosion as follows: Suspension, saltation and surface creep. These distinct types of movement usually occur simultaneously and take place near the surface at height not greater than approximately 1m.

Mechanism of wind erosion

Analysis is required on the mode and magnitude of the forces involved as they react upon soil particles. Three stages are involved in wind erosion mechanism:

b. Initiation of movement
c. Transportation
d. Deposition

Factors influencing the intensity of wind erosion

Wind erosion occurs whenever conditions are favorable for detachment and transportation of soil material by wind. Five factors influence the intensity of soil erosion due to wind activity:

1. Soil erodibility factor,
2. Soil surface roughness,

3. Climatic conditions (wind velocity and humidity), little can be done to change the climate of an area but it is usually possible to alter one or more of the other factors to reduce erosion.
4. Length of exposed surface,
5. Vegetative cover

Estimating wind erosion losses

Equations have been developed to indicate the relationship between the amount of wind erosion and the various field and climatic factors that influence erosion. This is used as a guide for determining

(a) The potential amount of wind erosion on any field under existing local climatic conditions.
(b) The conditions of surface roughness, soil cloddiness, vegetative cover, sheltering or width orientation of field.

The equation is being expressed as:

$$E = f(I', C', K', L'V) \dots\dots\dots\dots\dots\dots\dots\dots\dots.1.1$$

Where:
   E = Weight of annual erosion per unit area
   I' = A soil erodibility index
   C' = A climatic factor
   K' = A soil ridge roughness factor
   C' = Equivalent field length along the prevailing wind direction
   V' = Equivalent quality of vegetative cover

2.2.3 Soil compaction

Agricultural activities such as tillage practice causes some significant effect on soil compaction such as poor root development, poor soil aeration, poor soil water movement and development of hard pan in the soil. Soil compaction process occurs in tropical conditions due to:

a. Tillage and harvest operations carried out under wetter than the optimal conditions required for wheel movement,
b. Due to the excessive trampling of the draughts animal such as cattle etc in pasture areas and,

c. Due to the traffic of the harvest operations and transportation under inadequate soil water conditions in forest areas.

Interactions between tillage operations and soil compaction

Soil compaction is one of the reasons for the need for tillage. There are a number of interactions between tillage operations and soil compaction. Obviously, soil compaction may occur during a tillage operation as have been reported by several researchers. During tillage, a part of the soil is broken up into various size clods by the implements and a part is compacted by traffic (traffic-induced compaction).

Soil compaction from the engineering construction perspective is defined as the method of mechanically increasing the density of soil. In construction, this is a significant part of the building process. If performed improperly, settlement of the soil could occur and result in unnecessary maintenance costs or structure failure. Almost all types of building sites and construction projects utilize mechanical compaction techniques.

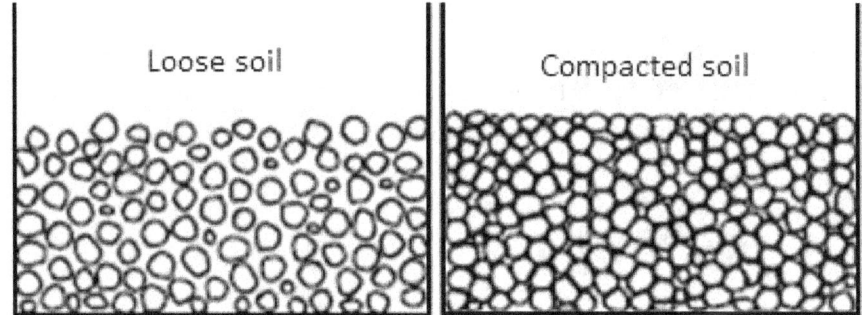

Figure 2-10: Soil density (*loose*, poor load support; *compacted*, improved load support)

Soil density estimation

Soil density test: To determine if proper soil compaction is achieved for any specific construction application, several methods were developed. The most prominent of all the tests is soil density test.

Why the test: Soil testing accomplishes the following:

a. Measures density of soil for comparing the degree of compaction vs specifications
b. Measures the effect of moisture on soil density vs specifications
c. Provides a moisture density curve identifying optimum moisture

Types of soil density tests

Tests to determine optimum moisture content are done in the laboratory. The most common is the Proctor Test, or Modified Proctor Test. A particular soil needs to have an ideal (or optimum) amount of moisture to achieve maximum density. This is important not only for durability, but will save money because less compaction effort is needed to achieve the desired results.

Proctor test (ASTM D1557-91)

The Proctor, or modified proctor test, determines the maximum density of a soil needed for a specific job site. The test is carried out to determine

a. The maximum density achievable for the materials and uses this figure as a reference. Secondly,
b. To test the effects of moisture on soil density.

The soil reference value is expressed as a percentage of density. These values are determined before any compaction takes place to develop the compaction specifications. Modified Proctor values are higher because they take into account higher densities needed for certain types of construction projects.

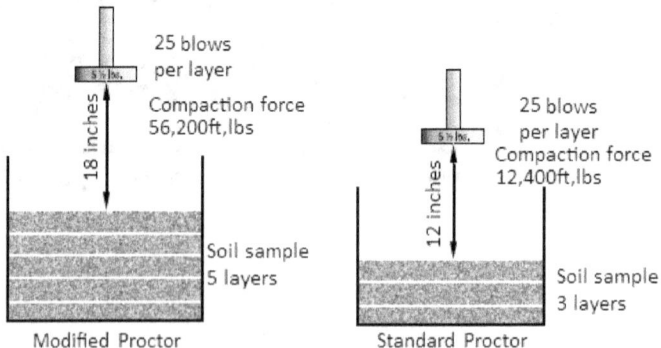

Figure 2-11: Proctor tests

Field tests: Sand cone test (ASTM D1556-90)

Sand cone test is a common field test used to determine if compaction densities are being reached on the spot. A small opening (6" x 6" deep) is dug in the compacted material to be tested. The soil is removed and weighed, then dried and weighed again to determine its moisture content. A soil's moisture is figured as a percentage.

The specific volume of the hole is determined by filling it with calibrated dry sand from a jar and cone device. The dry weight of the soil removed is divided by the volume of sand needed to fill the opening. This gives us the density of the compacted soil in lbs per cubic foot. This density is compared to the maximum Proctor density obtained earlier, which gives us the relative density of the soil that was just compacted (Figure 2-12).

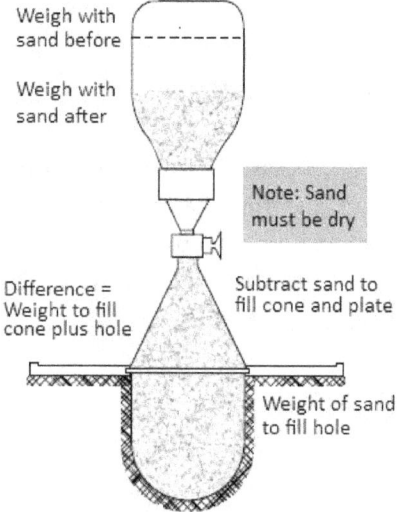

Figure 2-12: Sand cone test

Advantages of cone test include *its* ability to test large sample and accuracy.

Disadvantages include and it involves many steps as well as large area requirement, slow process among others.

Common errors associated with this test include presence of void under plate, tendency of sand bulking, compacted sand, and soil pumping

Soil modulus (soil stiffness)

This field-test method is a very recent development that replaces soil density testing. Soil stiffness is the ratio of force-to-displacement. Testing is done by a machine that sends vibrations into the soil and then measures the deflection of the soil from the vibrations. This is a very fast, safe method of testing soil stiffness. Soil stiffness is the desired engineering property, not just dry density and water content. This method is currently being researched and tested by the highway construction administrators.

2.3 Soil conservation practices

From the foregoing, soil resources are continually being depleted and their losses is an indication that the expanding population in many parts of the world is pressing this resource to limits. The solution to these problems is what is referred to as conservation practices which include the following; erosion control, erosion prevention and soil compaction measures.

2.3.1 Erosion prevention measures

Soil is not only lost in erosion process, but also a proportionally higher percentage of plant nutrient, organic matter and fine soil particles is lost. Erosion prevention is more economical than protection with structural measures. The control of soil erosion caused by water or wind is of great importance in the maintenance of crop yields. Thus in erosion prevention, emphasis should be given to the followings:

1. *Proper land management practices*

These measures include

a. *Practice of minimum tillage*: Minimum tillage encompasses a wide range of techniques, from direct drilling of seed into stubble or pasture, through spraying followed by direct drilling, or spraying followed by a reduced number of cultivation passes before sowing, to more judicious use of conventional ploughs and harrows.

Figure 2-13: Zero tillage conservation planting

No-till farming is a typical minimum tillage practice employed for growing crops without disturbing it through tillage. The process of tilling is beneficial in mixing fertilizers in the soil, making rows and preparing the surface for sowing. But the tilling activity can lead to compaction of soil, loss of organic matter in the soil and the death of soil organisms. No-till farming is a way to prevent the soil from this harm.

b. *Stubble mulching*: On continuously cropped soils, stubble can be mulched after harvest and left on the paddock if it is to be direct-drilled the next season. If it is to be cultivated, mulched stubble can be incorporated into topsoil, in the course of cultivation. In both circumstances, a partial ground cover of mulch protects soil from being lifted by wind or runoff until the new crop emerges. If soil particles are detached, most of them are caught in the mulch before they move far.

Figure 2-14: Stubble mulching

c. *Contour ploughing:* Where sloping ground is cropped, it can be cultivated along the contour as an alternative to the standard practice of cultivating up and down slopes. During rain, soil washed off ridges is trapped in furrows running across-slope, instead of being carried away through down slope furrows by runoff.

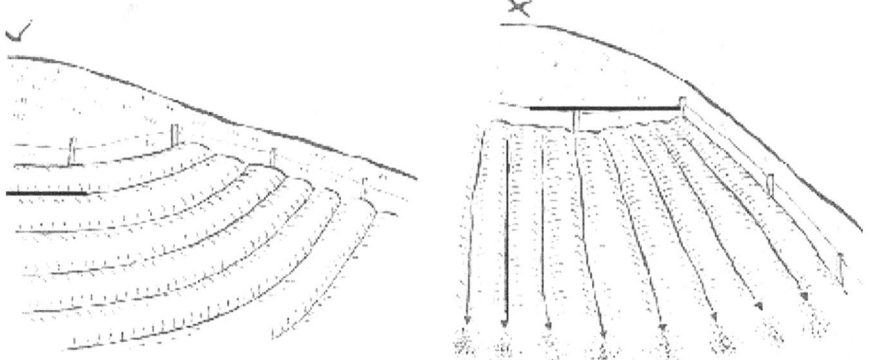

Figure 2-15: Contour ploughing

d. *Controlled grazing and revegetation of open grass ways*: As a result of animal grazing or heavy rainfall on cropland, surface runoff carries soil into depressions which act as temporary waterways. If they are bare, the concentrated flow washes the soil particles out of the field and into watercourses. Retaining grass cover provides resistance to water flow, slowing it to the point where it drops sediment.

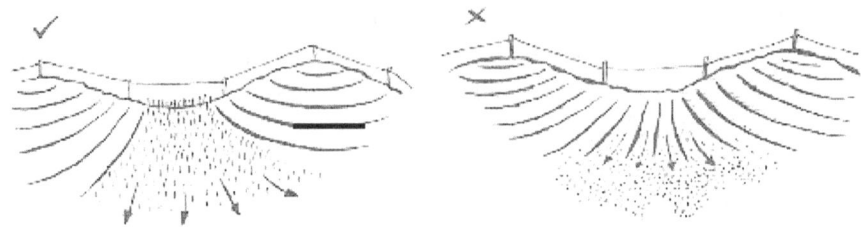

Figure 2-16: Grassed waterways

e. *Construction of windbreaks*: windbreaks have been planted for many years. Although their primary role is shelter, they are also used to protect soil against wind erosion. They are known to increase crop and pasture growth by about 10%, with other benefits being better stock condition and reduced deaths during cold weather, but these are consequences of windbreaks' shelter role, not erosion control.

Figure 2-17: Single row windbreak

Benefits of windbreak

- Prevention of forest fire and indiscriminate logging
- Control of road construction and mining activities
- Immediate stabilization of erosion.

2. *Bioethical measures*

Bioethical measures for erosion control include combination of simple structural mechanical and vegetative methods for erosion prevention. Examples include the interplanting of the riprap with vegetation as well as the use of prefabricated, heavy, galvanized wire baskets filled with rocks (gabions) at the bottom of a slope. An ingenious and economical method involves the use of worn-out automobile tires to construct retaining walls or to place them flat on the slopes (Figure 2-18).

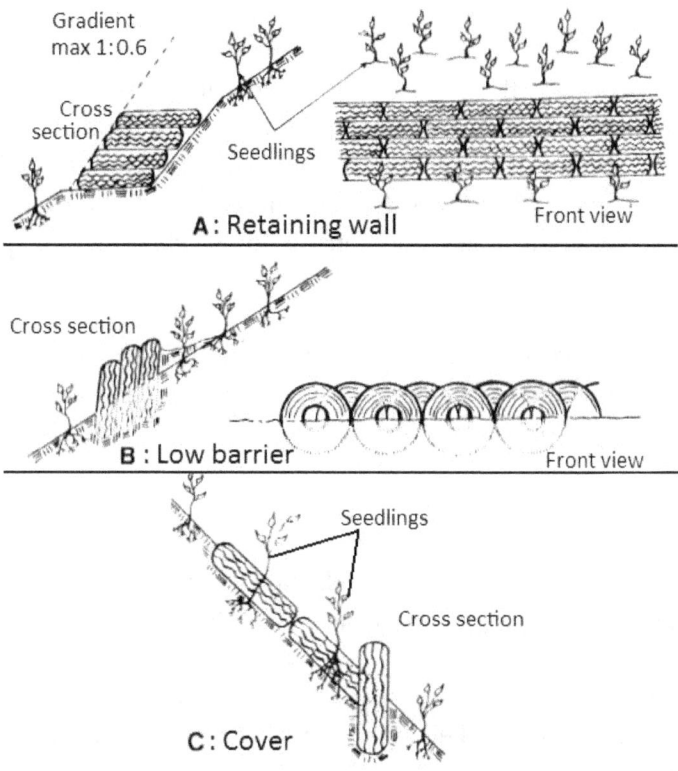

Figure 2-18: Use of worn tyres for stabilizing slopes (Weidelt, 1975)

3. *Physical soil conservation measures*

Physical soil conservation measures on cropland are generally engineering work based. Various structures are built using earth, stone or masonry to reduce the effect of slope length and to intercept and slow down run-off water. Conservation structures are usually made through human labour, or more rarely with machinery.

The main physical conservation measures on cropland are:

a. Bunding,
b. Terracing,
c. Creating diversions (cut-off drains),
d. Vegetated waterways and
e. Physical work for gully control.

These are important control measures often necessary on range and pasture lands and in forest areas. Simple engineering knowledge is required in terraces, waterways and laying-out of contours.

a. Bunding

Bunding and terracing are the most common engineering practices employed in soil conservation. Bunds are ridges or embankments or long and narrow projections constricted on the surface on the land at selected places and in selected directions. The bunds may be of different sizes and heights and constructed for various purposes. People followed these methods traditionally to develop their sloping lands to conserve the soil for cultivation.

Structure of a bund: The basic structure of the bund is trapezoidal. It has a wider bottom, narrow top width, two sloping sides and a height. The sides are equal when .the bund is constructed on a plain ground.

Types of bunds

Depending on the purpose the bunds are classified as contour bunds, graded bunds, side bunds, lateral bunds, marginal bunds and shoulder bunds.

i. *Contour bunds*: When bunds are constructed across the slope in a contour line they are called contour bunds. They are constructed to reduce the length of the slope by dividing slope into different sections and to impound the running water at different sections to increase percolation. They are normally constructed on lands with 1-6 per cent slope. But bunds may be constructed up to eight per cent slope.

Figure 2-19: Contour bund

ii. *Graded bunds*: When the contour bunds are made with slight deviation from the contour line towards one direction to drain out the surplus water that may occur during the rain, they are called graded bunds. For a casual observer the graded bunds look like contour bunds. Graded bunds are made on land with slope between 2-8 per cent.

Figure 2-20: Graded bund

iii. *Side bunds:* Side bunds are the bunds that are constructed at the extreme ends of the contour bunds along the line of slope. But in the case of graded bunds the side bund is constructed as part of the drainage channel on the side towards which the water is drained. But side bunds are constructed on both sides of the contour and graded bunds.

Figure 2-21: Side bund

iv. *Lateral bunds:* Bunds constructed along the slope between two side bunds in order to prevent concentration of water along one side and to break the length of the contour bund into convenient bits are called lateral bunds. Usually the lateral bunds are constructed when the contour or graded bunds are more than 300 meters long. Like the side bunds the lateral bunds also should be accompanied by drainage channel to drain out the excess of rain water.

Figure 2-22: Lateral bund

v. *Marginal bunds:* Bunds constructed along margins of the watershed, boundaries, road margins, river or stream margins, gully margins and the like are called marginal bunds. They serve as boundary bunds.

Figure 2-23: A road over marginal bund

vi. *Shoulder bunds:* Shoulder bunds are those constructed at the outer edge of a terraced plot. When a terraced plot is used for wet paddy cultivation it is bunded on the outer side in order to retain the water. Sometimes shoulder bunds are constructed in order to prevent the rain water flow out of the terraced plot and thereby allow more percolation into the plot.

Figure 2-24: Shoulder bund

vii. *Stone bunds:* Stone bunds are constructed on steep farmlands where the soil is shallow and large quantities of stones are available. They are usually constructed in areas where there is a high population density and a shortage of arable land, resulting in steep land with shallow soils being cultivated.

Figure 2-25: Stone bund stabilized with grass

The bund, which should be about 0.5 m wide and 0.75 m high, is made of stones laid in ridges along the contour. The vertical interval between consecutive terraces is similar to the other terraces. If well maintained, terraces will gradually be formed due to the accumulation of soil above the bund.

b. *Terracing*

Soil conservation terraces on croplands are artificial earth or stone embankments, or combined channels and embankments, constructed across sloping land at a fixed or calculated vertical interval down the slope. Terracing is the second most commonly used method for soil conservation cropping land. Terrace is a cut and leveled portion of land on a slope. The important effect of terraces is to reduce the erosive force of the run-off water. Terraces are classified on the basis of width, slope and the material used for the construction of the terrace wall. Typical terraces commonly in use for erosion control; channel terraces, bench terraces contour terraces, and parallel terraces.

i. *Channel terraces*: Digging a trench along the contour and throwing the soil on the lower side makes a channel terrace. If the embankment below the channel is narrow, the resulting structure is called a narrow-based channel terrace

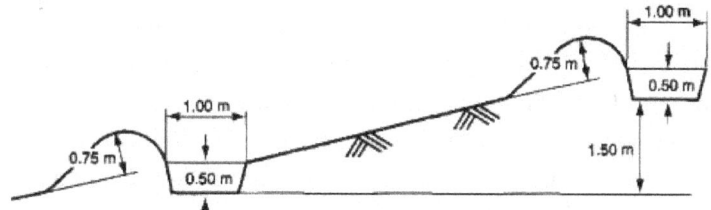

Figure 2-26: Channel terraces with a vertical interval of 1.50 m

ii. *Bench terraces*: These are level or nearly level terraces constructed on the contour and separated by embankments (risers). They can be formed by excavation or may develop over time from a grass strip, hedges, trash-lines, or channel terraces. They are mainly constructed on relatively steep slopes (more than 35% slopes) that have to be cultivated due to shortage of flat or gently sloping arable land.

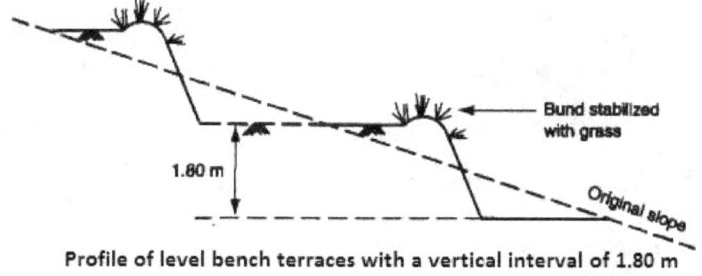

Profile of level bench terraces with a vertical interval of 1.80 m

Figure 2-27: Bench terrace

Bench terraces reduce land slope and allow run-off from the upper side of the terrace to go into a lower portion where it spreads out and infiltrates.

Figure 2--28: Rice terracing

iii. *Contour terraces* have point rows and grassed waterway outlets that follow the lay of the land. This practice of farming on slopes takes into account the slope gradient and the elevation of soil along the slope. It is the method of ploughing across the contour lines of a slope. This method helps in slowing the water runoff and prevents soil from being washed away along the slope. Contour ploughing also helps in percolation of water in the soil.

iv. *Parallel terraces* eliminate the production losses associated with point rows and minimize the interference to farming operations when spaced at multiple widths of planting and harvesting equipment. The width of the terrace depends on the degree or the percentage of slope. The greater the slope the lesser will be the width of the terrace for the same height of the cut portion to make the terrace.

c. *Diversions (cut-off drains) ditches*

A diversion ditch is a graded channel designed to intercept surface run-off and convey it safely to an outlet or waterway or, if so desired, into a farm (in low-rainfall areas). A diversion ditch usually has a larger capacity and steeper gradient than a terrace channel.

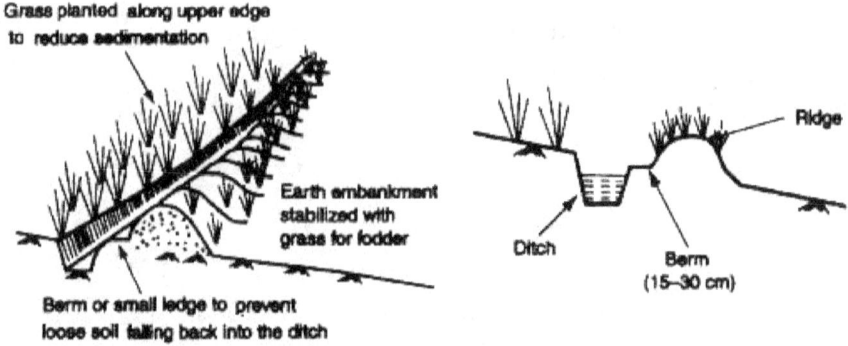

Figure 2-29: A diversion ditch

When diverting water from a diversion ditch, the outlet point should be safe from erosion so that a new gully will not be formed.

d. *Waterways*

Waterways are needed to conduct run-off safely from hill slopes to valley bottoms where it can join a stream or river. Where there is a natural depression or small valley that is well stabilized with vegetation this may be adequate to take the discharge from diversion ditches or graded terraces, but where there is no such natural waterway, an artificial waterway (drainage way) must be installed.

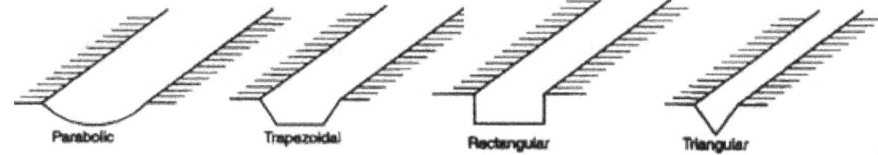

Figure 2-30: Different cross-section shapes for waterways

Artificial waterways are of various shapes (e.g. trapezoidal, parabolic, rectangular or triangular; Figure 2-31) depending on the discharge expected and the material used for stabilization, i.e. whether grass, stone, masonry or concrete (Figure 6.10). Grassed waterways are usually trapezoidal or parabolic and wide and shallow to allow water to flow in a thin sheet. Where slopes are steep, grassed waterways should be made wide enough to spread the water and prevent high velocities of the water.

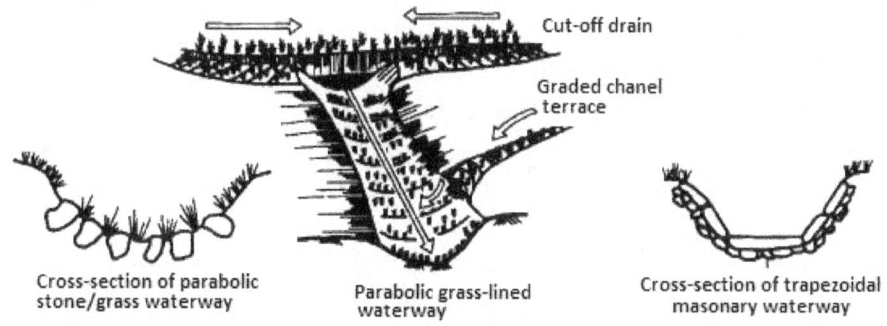

Figure 2-31: Different types of waterway

e. *Physical work for gully control*

Gully control is a very expensive undertaking, and it is always better to prevent gullies from being formed in the first place. Prevention of gullies is certainly more economical than cure. If there are several gullies, and if it is difficult to find an outlet point for discharged water, one of the larger gullies should be used as a drainage way.

When a medium-sized or large gully (>1 m deep) already exists, the best that can be done is to try and halt continued erosion by construction of permanent structures combined with protection from grazing and planting of grass, shrubs and trees.

2.3.2 Erosion control measures

Three major erosion control measures (water or wind erosion) includes; Contouring, strip cropping, and tillage practices.

Water erosion control measures

For effective water erosion control and better crop use, the following erosion control practices are recommended where applicable:

1. Performing all planting, tillage and harvesting operations on or nearly on the contour.
2. Plant close growing and intertilled crops in alternate strip
3. Construct cross shape channel (terraces) to carry water off at reduces velocities.
4. Planting belts or trees or constructing other barriers for protection from wind erosion.
5. Using crop residues on the surface or incorporated in the top soil with different tillage methods.
6. Performing minimum tillage operation consistent with good weed control and seed bed preparation.
7. Establishing permanent vegetation in watering and other eroded areas.
8. Stabilizing gullies and wasteland areas with suitable structures.

Wind erosion control measures

The two major types of wind erosion control measures consist of those measures that reduce surface wind velocities and those that affect soil characteristics. Three principal methods of reducing wind surface velocities are:

1. *Vegetative measures*; measures such as cultivated crops, trees and shrubs etc. can be used to control soil erosion.
2. *Tillage practices*: Soil erosion can be minimised by adopting conservation farming practices that maximise retention of crop residue and reduce pulverisation of soil.

Figure 2-32: Surface roughening and retention of stubble (John Lays, 2003)

3. *Mechanical and binding methods* such as wind breaks, slat or brush fences, board walls, paper strip as well as surface protection (brush matting, rock and gravel). These methods tend not to be effective on sandy soils or are too expensive to be used on large areas (more than 10ha) because sandy soil has little soil structure.

Figure 2-33: Mechanical and binding methods

Many measures such as permanent grassing and contouring provide both types of control. For instance vegetation retards surface wind and also increases soil aggregation in fine textured soils.

2.3.3 Practical solutions to reduce soil compaction

The risk of subsoil compaction can be reduced by several measures as presented below in the order of importance

1. The easiest way to avoid soil compaction, at least theoretically, is to avoid field operations under soil conditions that are susceptible to compaction. This would mean avoiding field operations when the subsoil is wet.

2. Conventional ploughing can be replaced by on-land ploughing (or of course, by reduced tillage). On-land ploughing considerably reduces the risk of subsoil compaction compared with conventional ploughing.

3. To avoid soil compaction, the harvester could be unloaded at the head of the field. This is of course only possible in practice when the capacity of the harvester is adjusted to the dimensions of the field or vice versa. If transport traffic cannot be avoided, then at least the tractors and trailers used for such transport should have good tyre equipment.

4. Another potential but unpopular way to reduce contact stress is to use tracks instead of wheels. Tracked tractors usually have a greater contact area than wheeled tractors with equivalent power ratings. Rubber tracks have a great potential to provide better tractive performance than wheels either over a wider range of drawbar pull with a tractor of the same weight or over the same range with a much lighter tractor (Okello et al., 1994). The stress distribution below the tracks may be very uneven, resulting in high maximum stress and hence a high risk of compaction. However, in cases where the rubber-tracked tractor is well-balanced (the maximum stress is as small as possible) the risk of subsoil compaction may be lower than with wheeled tractors of similar size.

5. Controlled traffic is a system to reduce compaction by limiting compaction to designated areas of the field, i.e. the wheel tracks or 'tramlines'. All machinery is then adjusted to e.g. 3 m wheel base and 9 m implement width, while the width of the tyres or tracks is kept relatively small. While this concept is very appealing, it is probably limited to large farms that have more or less rectangular fields. It is also difficult to apply in tillage systems that include mouldboard ploughing.

2.4   Water (soil moisture) conservation practice

Deficiency of soil moisture for crop is a major physical problem of agriculture. These problems are most serious essentially in the Northern part of Nigeria where rainfall distribution is not uniform in the growing season. To solve this perennial problem, the agricultural engineers and other scientists have humorous duty in developing new practices that will permit storage of greater percentage of the available precipitation in the soil profile.

Typical of the challenges facing engineers in solving these problems among other things include

1. Tillage practices that modifies the soil surface configuration so as to refrain precipitation
2. Reduce the total evaporation potential on soil surface

3. Design of effective level bench terrace systems with special water catchments areas and
4. Surface evaporation control through the use of mulching and films.

Soil types and conditions

Soil types are commonly classified by grain size, determined by passing the soil through a series of sieves to screen or separate the different grain sizes (Figure 2-34). Every soil type behaves differently with respect to maximum density and optimum moisture. Therefore, each soil type has its own unique requirements and controls both in the field and for testing purposes.

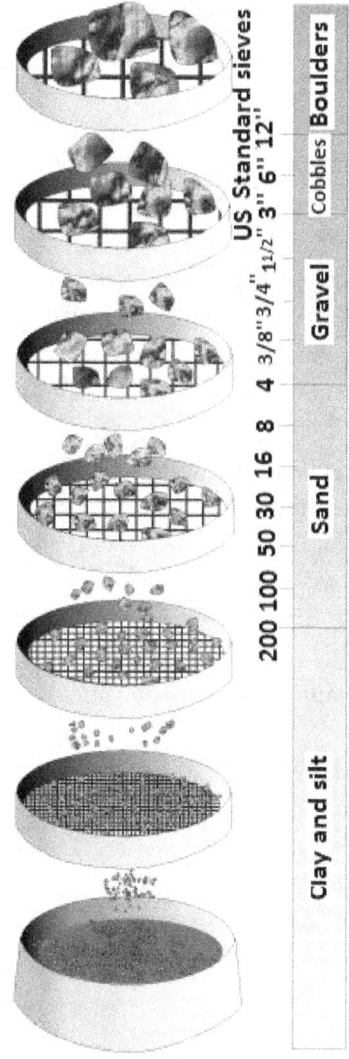

Figure 2-34: Soil grain sizes

Soil is categorized by a system set up by American Association of State Highway and Transportation Officials (AASHTO). Soils found in nature are almost always a combination of soil types. A *well-graded* soil consists of a wide range of particle sizes with the smaller particles filling voids between larger particles. The result is a dense structure that lends itself well to compaction. A soil's makeup determines the best compaction method to use.

There are three basic soil groups:

1. *Cohesive soils:* Cohesive soils have the smallest particles. Clay has a particle size range of 0.00004" to 0.002". Silt ranges from 0.0002" to 0.003". Clay is used in embankment fills and retaining pond beds.

   Characteristics:

   a. Cohesive soils are dense and tightly bound together by molecular attraction.
   b. They are plastic when wet and can be moulded, but become very hard when dry.
   c. Proper water content, evenly distributed, is critical for proper compaction. Cohesive soils usually require a force such as impact or pressure.
   d. Silt has a noticeably lower cohesion than clay. However, silt is still heavily reliant on water content.

2. *Granular soils:* Granular soils range in particle size from 0.003" to 0.08" (sand) and 0.08" to 1.0" (fine to medium gravel). Granular soils are known for their water-draining properties.

   Characteristics

   a. Sand and gravel obtain maximum density in either a fully dry or saturated state.
   b. Testing curves are relatively flat so density can be obtained regardless of water content.

3. *Organic soil:* Organic soils have low permeability, poor foundation support and are generally not suitable for compaction.

Table 2-2: Summary of soil characteristics

Fill materials	Properties				
	Permeability	Foundation support	Permanent subgrade	Expansive	Composition difficulty
Gravel	Very high	Excellent	Excellent	No	Very easy
Sand	Medium	Good	Good	No	Easy
Silt	Medium low	Poor	Poor	Some	Some
Clay	None	Moderate	Poor	Difficult	Very difficult
Organic	Low	Very poor	Not acceptable	Some	Very difficult

Source: Multiquip, 2012

2.5 Soil and water relationship

The knowledge of soil and water relationship is valuable to irrigation practice. These relationships include soil bearing capacity, the flow and movement of water in the soil, the salinity and alkalinity of the soil, translocation and concentration of salts due to movement and evaporation of the soil.

Excessive water volume in the soil retards or inhibits plant growth and lack of water in the soil has been responsible for sterility of arid region. The behaviour of water in the soil is due to certain physical characteristics of the soil such as its structure, texture, water holding capacity etc.

Soil moisture

The response of soil to moisture is very important, as the soil must carry the load year-round. Rain, for example, may transform soil into a plastic state or even into a liquid. In this state, soil has very little or no load-bearing ability.

Moisture regimes of the soil

1. *Saturation*: A state in which all available pore space is filled with water. When soil is saturated, all void spaces are filled up.
2. *Field capacity*: The state of water in the soil after that part of it draining has left. This is water left after 2 days of gravitational water has drained off leaving the soil at field capacity.

3. *Wilting point*: As plants draws water out of the capillary pores, readily at first and then with greater difficulty, until no more water can be withdrawn and the only water left is in the micro-pores. The soil is then at wilting point and without water additions, plants will die

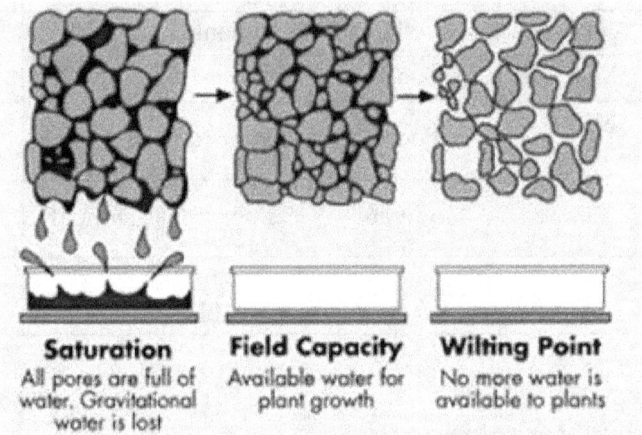

Saturation
All pores are full of water. Gravitational water is lost

Field Capacity
Available water for plant growth

Wilting Point
No more water is available to plants

Figure 2-35: Stages of water holding capacity Source: BETTER SOILS

4. *Permanent wilting point*: The moisture regime of the soil when a plant permanently wilts. Field estimates of PWP are determined by measuring the soil moisture content at which plant in it have permanently wilted. Tension at PWP varies from 7.0 – 40 atm.

5. *Available water*: The moisture content of the soil between the field capacity and the permanent wilting point. The amount of soil water available to plants is governed by the depth of soil that roots can explore (the root zone) and the nature of the soil material. Because the total and available moisture storage capacities are linked to porosity, the particle sizes (texture) and the arrangement of particles (structure) are the critical factors.

6. *Readily available water* – that part of available water where not much energy is exerted to get it to the plant. This is approximately 75% of the total available water

Table 2-3: Water holding capacity (mm/cm depth of soil) of main texture groups.

Texture	Field capacity	Wilting point	Available water
Coarse sand	0.6	0.2	0.4
Fine sand	1.0	0.4	0.6
Loamy sand	1.4	0.6	0.8
Sandy loam	2.0	0.8	1.2
Light sandy clay loam	2.3	1.0	1.3
Loam	2.7	1.2	1.5

Sandy clay loam	2.8	1.3	1.5
Clay loam	3.2	1.4	1.8
Clay	4.0	2.5	1.5
Self-mulching clay	4.5	2.5	2.0

[1]*Source: BETTER SOILS (n.y.)*

Classes of soil water

1. *Hygroscopic water* – water on the surface of the soil grains and is not capable of movement by action of gravity or capillary force.
2. *Capillary water* – This is that part in excess of hygroscopic water which exists in the pore spaces of the soil and is retained against gravity.
3. *Gravitational water* – This is the part in excess of hygroscopic and capillary water which move out of the soil if favourable drainage is provided.

Estimating water available to plants

Soil moisture measurement methods include:

Direct measurement

a. *Feel method*: A quick method of determining moisture density is known as the "*hand test.*" Pick up a handful of soil. Squeeze it in your hand. Open your hand. If the soil is powdery and will not retain the shape made by your hand, it is too dry. If it shatters when dropped, it is too dry.

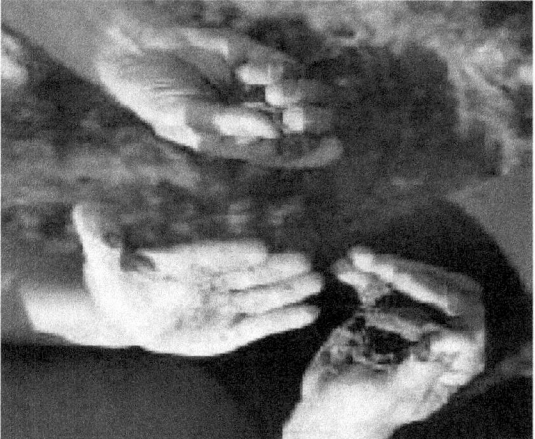

Figure 2-36: Moisture test by feeling

---

[1] Figures are averages and vary with structure and organic matter differences.

Table 2-3: Descriptions are for sandy loam soils

Moisture content	Description
0–25%	Ball of soil is weak and grains break away quickly
25–50%	Soil ball is weak, but finger marks show
50–75%	Moist ball, dark colour, light soil stain on fingers
75–100%	Moist soil ball, medium soil stain on fingers

If the soil is mouldable and breaks into only a couple of pieces when dropped, it has the right amount of moisture for proper compaction. If the soil is plastic in your hand, leaves small traces of moisture on your fingers and stays in one piece when dropped, it has too much moisture for compaction.

Indirect measurements

This is done with the use if instrumentation such as:

a.   *Tensiometer:* A tensiometer is a water-filled tube with a special porous tip and a vacuum gauge used in measuring soil water suction, which is similar to the process a plant root uses to obtain water from the soil.

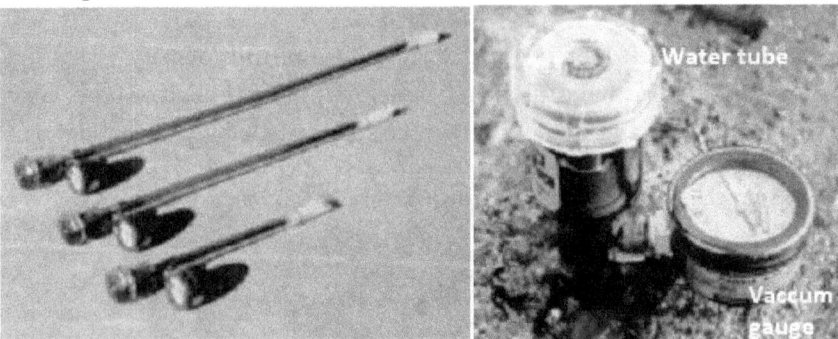

Figure 2-37: Tensiometers

Tensiometers are available in various lengths as pictured at left. When installed, only the top of the water-filled tube and the vacuum gauge are visible for readings and field maintenance as shown at right.

b.   *Gypsum blocks*: Gypsum blocks consist of two electrodes embedded in a block of gypsum. They can be either cylindrical or rectangular in shape with concentric or parallel electrodes (Figure 1). Wires are joined onto each electrode and extruded from the gypsum block to measure the resistance between the electrodes. The resistance between the two electrodes varies with the water content in the gypsum block, which will depend directly on the soil water tension.

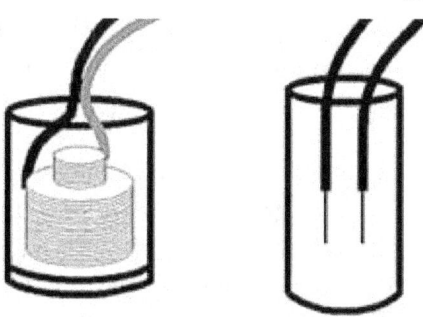

Figure 2-38: Internal arrangement of electrodes in cylindrical gypsum blocks

As the soil dries out water is extracted from the gypsum block and the resistance between the electrodes increases. Conversely as the soil wets, water is drawn back into the gypsum block and the resistance decreases.

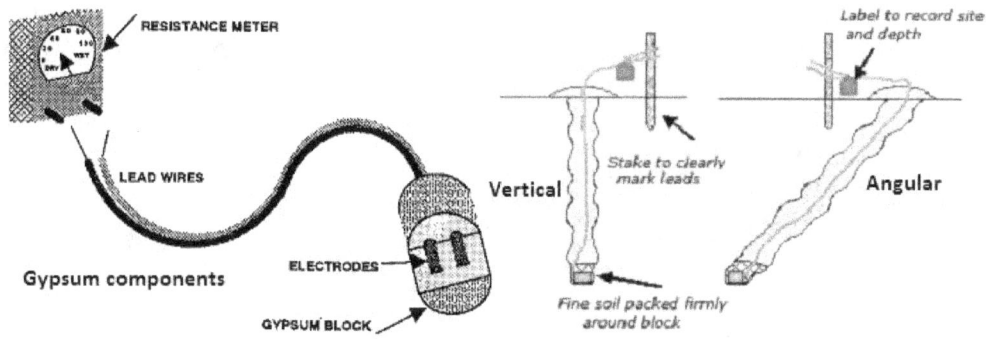

Figure 2-39: Installation methods for gypsum blocks

c. *Lysimeters*: Lysimeters are buried containers of soil equipped with a weighing device and drainage system to measure evaporatranspiration and percolation. Lysimeters aim to represent existing soil, vegetation, and climatic conditions to improve the accuracy of measurements of physical processes (Hillel, 1998).

Figure 2-40: Installation methods for gypsum blocks

A lysimeter is most accurate when vegetation is grown in a large soil tank which allows the rainfall input The two different types of lysimeter are in common use; weighing and nonweighing lysimeters.

d. *Soil probe*: A soil probe or auger is the best tool for sampling soil. An auger will be needed if the soil is very stony or gravelly. Simply push the probe (or push and turn the auger) into the soil to the desired depth, lift up to remove the core, and place it in the clean pail. Sampling depth should be 4 to 6 inches deep for lawns, turf, or other perennial sod, or tillage depth (usually 6-10 inches) for annually tilled crops. For proper soil sampling, the soil probe should be inserted into the soil perpendicular to the ground, and to the appropriate depth.

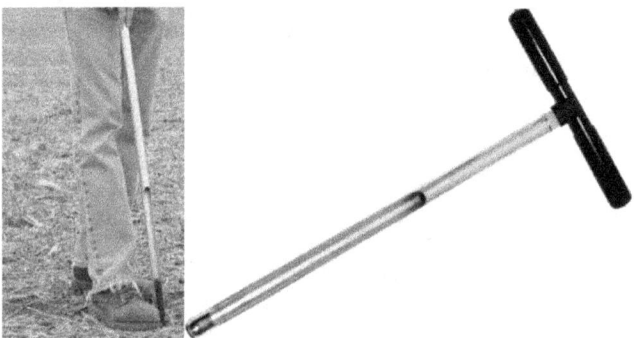

Figure 2-41: Soil probe

e. *Soil moisture sensors*: Moisture sensors measure the water content in soil. A soil moisture probe is made up of multiple soil moisture sensors.

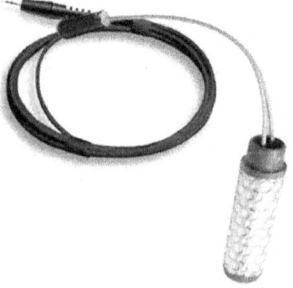

Figure 2-42: Typical soil moisture sensor

f. *Rooting depth methods.* For determining the water losses due to deep percolation and water uptake by a crop, the depth of rooting must first be known. The depth of root system varies from crop to crop and also from time to time during growth. The root system is influenced by several factors; too much wetness results in development of a shallow root system while drier soil water regimes encourage a deep one; soil texture and structure influence the depth of the root system to a great extent. Studies on rooting depths should, therefore, be made under existing and recommended cultural and irrigation practices on representative soils. The two most common methods of measuring the rooting depth are by excavation method and by studying soil moisture extraction patterns. rooting depth of crop plants. Components are measured directly or indirectly and either individually or in an integrated way.

Moisture vs. soil density

Moisture content of the soil is vital to proper compaction. Moisture acts as a lubricant within soil, sliding the particles together. Too little moisture means inadequate compaction—the particles cannot move past each other to achieve density. Too much moisture leaves water-filled voids and subsequently weakens the load-bearing ability. The highest density for most soils is at a certain water content for a given compaction effort. The drier the soil, the more resistant it is to compaction.

2.6   Soil conservation equipment

Measures taken to preserve soil water prevent excessive loss of soil to erosion and drainage. The following equipments are essential for large scale soil conservation projects:

a. *Ceaders* - for making bunds across the slope to prevent uncontrolled flood flowing down hill.
b. *Ditchers*-This equipment is used for making drains and waterways for erosion control and drainage.
c. *Ridger* – These are disc implements used for making big ridges and bunds for erosion control. They are characterized by two heavy duty discs with wide enough diameter to cut and throw soil in the same direction thereby piling up the soil along a border.

Figure 2-38: Two blade disk ridger

d. *Bunders* – Bunders are used for making border bunds. Light draft border disk plough builds high, well-shaped bunds and borders. Disks (28-inch diameter) turn easily on double tapered roller bearings that are pressure-lubricated. They differ from ridgers in weight and diameter of blade disc.

Figure 2-43: 3-blade border disk plough

e. *Excavators* - for making trenches or grassed waterways for drainage

Figure 2-44: Excavator

Further reading

Amy Vickers (2001) *"handbook of water use and conservation" water plow press,*

Beasly R. P., Gregory J. M., and McCarty T. R., (1984).Erosion and Sediment Pollution Control, 2d ed., (Ames: Iowa State University Press,), 242-43.

Fellenius W., 1979. *Calculation of the Stability of Earth Dams,* Trans. 2nd Congress on large Dams, Washington,.

John Lays, 2003. Wind erosion. Pub. Centre for Natural Resources NSW Department of Infrastructure, Planning and Natural Resources Parramatta. September 2003 ISBN 0 7347 5399 3 CNR 2003.069

Multiquip, 2012. Soil Compaction: a basic handbook. mq_soil_handbook.pdf. date assessed: 04/04/2012

Ozara N.A and Madubuike C.N. (1994). Biotechnical measures for erosion control on agricultural fields. Proceedings of the 1st S.E. Na/da Pre-season workshop organized under NALDA NAERLS memorandum of understanding.

SCS (Soil Conservation Service, USDA), *1975a.* Engineering field manual. Washington, DC: USDA.

Texas Agricultural Extension Service (2003). Agricultural water conservation practices. *Www.twdb.state.tx.us/assistance/conservation/agricons.htm*

Van Morrill and Gabrielle Belfit (Eds.) (1999) CAPE COD-A Community Connected By Water. Part One, Water Resource Education Information brochure, November 1999

Weidelt, H. j., 1975 (compiler). Manual of reforestation and erosion control for the Philippines. Eschborn , West Germany: German Agency for Tech. Coop. Ltd.

Welter H. Wischmier and Dwight D. Smith (1976) predicting rainfall erosion losses. A guide to conservation planning US department of agriculture, Agriculture handwork No 537.

CHAPTER 3

Principles of Irrigation and Drainage

3.   Introduction

The engineering problem involved in water conservation can be divided into the following three phases.

1. Crop water supply (Irrigation)
2. Removal of excess water and
3. Flood control

Each of these engineering problems are discussed in the following sections

3.1   Crop water supply (irrigation)

Irrigation is defined as the application of water to land using means other than the natural rain, the purpose of which is to provide sufficient water for plant growth and productivity. Irrigation is necessary to provide enough water to fill the deficit arising from the depletion of soil moisture from the combine action of two separate phenomena of evaporation and transpiration.

Irrigation provides one of the greatest possibilities for increased production. Although it is fairly practiced in the Northern Nigeria, the south is yet to embrace it with zeal and vigor.

On-farm water use can be reduced substantially without decreasing productivity through improved irrigation technologies and efficient water management practices. Accurate water measurement and soil moisture monitoring are key components of efficient on-farm water management practices.

Irrigation flow meters can be used to help calculate the efficiency of irrigation systems, identify water loss from leaks in conveyance systems, and to accurately apply only the necessary amount of water based on soil moisture levels and weather conditions.

Soil moisture monitoring is used in conjunction with weather data and crop evapotranspiration requirements to schedule irrigation. The agricultural engineers and other applied scientists has a major responsibility in helping to meet these problems through development and application of practices that will reduce the unnecessary losses in storage and transmission thus providing a higher percentage of water for actual use by the irrigated crop.

3.1.1 Sources and quality of irrigation water

Precipitation in form of rain is the major sources of all water. Other sources of irrigation water are:

1. Surface reservoirs either natural (lakes) or artificial (dams).
2. Groundwater through abstraction and storage or direct use.
3. Rain-making (cloud seeding) through introduction of silver iodide and dry ice (solid CO_2) into air to initiate formation of raindrops.
4. Saline water conversion through desalinization of sea water.
5. The success of any irrigation depends among other things on the quality of water used.

Water quality degradation

Irrigation degrades water quality through

- Discharge of excess or tail water into streams
- Water erosion of sediments from irrigated fields
- Leaching of contaminated water into aquifers

Agronomic impacts of degraded water quality

- Limits its reuse for irrigation
- Decreases crop yields
- Can shorten the life of irrigation equipment

Benefits of irrigation includes

1   It enables growing of crops where it would otherwise have been impossible due to lack of water e.g. arid region.
2   Planting of crop through the year round.

3 Greater employment opportunities especially for those engaged in the industry –
   equipment manufacture, practitioners in the field engineers etc.
4 Greater utilization of land area for agricultural production.

Problems of irrigation

There are many problems arising from insufficient use of water for irrigation practices
such as:

- Loss of water as conveyance losses
- Evaporation losses
- Seepage losses and other losses

3.1.2 Water requirement and irrigation efficiency

Water requirement (WR) is defined as the quantity of water regardless of its source
required by a crop in a given period of time for its normal growth under field
conditions. This includes losses due to Evapo-Transportation (ET) plus losses during
application of irrigation water and the quantity of water required for special operation
such as leaching etc.

$$WR = ET + L_A + S_n \text{ ... ...3.1}$$

Where
   WR = Water requirement
   ET = Evapotranspiration losses
   L_A = Application losses
   S_n = Special needs

Net irrigation requirement – The depth of irrigation water exclusive of precipitation
carried over soil moisture, groundwater contribution etc. that is required
consumptively for crop production. It is the amount of water required to bring the
soil water level in the effective root zone to field capacity.

$$d = n \sum_{i=1}^{n} \left(M_{fci} - M_{bi} \right) A_i D_i \text{ ... ...3.2}$$

Where
   d = Net amount of water to be applied during an irrigation in cm.
   M_{fci} = Field capacity moisture content in the ith layer of the soil in percentage.

A_i = Bulk density of soil in the ith layer

D_i = Depth of the ith soil layer cm within root zone and

n = Number of soil layers in the root zone D.

Gross irrigation requirement: This is the total amount of water applied through irrigation. It is the net irrigation requirement plus losses in water application and other losses.

$$G_{IR} = \frac{N_{IR}}{\eta} \dots\dots\dots\dots\dots\dots\dots\dots\dots\dots.3.3$$

Where

G_{IR} = Gross irrigation requirement

N_{IR} = Net irrigation requirement

η = Field efficiency of the system

Irrigation efficiency: This indicates how efficiently the available water supply is being used based on the different methods employed. In designs, the degree of land preparation, and the experience of the irrigator are the principal factors influencing irrigation efficiency.

Component of irrigation efficiency

1. *Water conveyance efficiency*: The measure of the efficiency of water conveyance in canal network, field channels etc.

$$\varepsilon_c = \frac{W_f}{W_d} x\ 100\% \dots\dots\dots\dots\dots\dots\dots\dots\dots\dots.3.4$$

Where

W_f = water delivered to the irrigated plot.

W_d = water diverted from the sources.

2. *Water application efficiency*: This is the measure of how efficiently the water is applied.

$$\varepsilon_a = \frac{W_s}{W_f} x\ 100\% \dots\dots\dots\dots\dots\dots\dots\dots\dots\dots.3.5$$

Where w_s = water stored in the root zone of plant.

Water application efficiency below 100% is due to:

- Seepage losses from field distribution channel
- Deep percolation losses
- Losses as evaporation in sprinkler system and surface system.

3. *Water distribution efficiency*: Uniform distribution of water over whole irrigated area. Permissible length of run in furrow irrigation is controlled to a large extent by the efficiency of water distribution.

$$\varepsilon_d = 1 - \frac{\check{y}}{d} x\ 100\% \dots \dots \dots \dots \dots \dots \dots \dots \dots ..3.6$$

Where

d = average depth of water stored along the run during irrigation

$\check{y}$ = average numerical deviation from d.

Canal and conveyance system management

Lining canals with concrete or other liners reduces water loss through seepage by 10-30%. Evaporation in canals can be reduced if irrigation districts provide water on demand rather than keeping the canals continuously filled. Using underground conveyance systems eliminates costly evaporation and deep percolation.

Figure 3-1: A field under irrigation

Irrigation scheduling

Irrigation scheduling involves managing the soil reservoir so that water is available when the plants need it. Soil moisture and weather monitoring are used to determine when to irrigate, and soil capacity and crop type are used to determine how much water should be applied during irrigation. Scheduling of irrigation is essential to attain high application efficiencies, to avoid deep percolation and runoff losses of water and chemicals, and to prevent stress of the crop.

Scheduling surface irrigation is difficult for three reasons:

1. It is often difficult to quantify the amount of water applied,
2. It is difficult to quantify the uniformity of the application, and
3. Water deliveries from irrigation water suppliers often are not under the farmer's control.

Irrigation management under variable rainfall conditions

In regions where rainfall may be significant, the effective rainfall must be considered to reduce irrigation requirements and minimize the negative effects of over-irrigation. These impacts include deep percolation, transport of nutrients and solutes into groundwater, surface runoff, and soil erosion. The scheduling of irrigation therefore requires data on potential future rainfall. This information can be in the form of a short-term weather forecast, or using information relative to previous rainfall records covering a significant number of years or using randomly generated rainfall events

Irrigation extension systems

The practical application of irrigation scheduling techniques requires appropriate technology transfer and support to farmers. *Irrigation extension or consulting services* provides this support and help in the transfer of technologies from research to practice.

Such services should not only provide information but also stress how this information is being used and what the impacts are for improving irrigation systems and management. De Jager and Kennedy defined three levels of technologies which must be adjusted to the technology level of the user (farmers) as follows:

1. *Top technology.* This situation means that there is a weather station in the farmland with a support system for decision making that allows individualized complete scheduling.
2. *Intermediate technology.* The station is strategically located for a group of farms that receive common information. This general information may be complemented individually with specific data regarding each field, such as soil water properties,

and irrigation systems. This information allows for the individual scheduling of irrigation. There is also a collective technical support system available.

3. *Low technology.* In this situation, fixed irrigation periods and volumes are scheduled for the whole campaign, based on large series of climatic parameters and crop average water requirements. Hill and Allen. (1996) present an interesting scheduling system for this situation.

There are many types of irrigation systems which usually fall into one of the three categories; *surface subsurface* and *overhead irrigation.* Surface irrigation systems are those that depend on gravity to spread the water across the surface of the land. These systems also are referred to as gravity or flood irrigation systems.

The shape of the soil surface and how the water is directed across the surface determined the types of surface systems (i.e. furrow, border, or basin). Sprinkler systems attempt to mimic rainfall by spraying the water evenly across the soil surface. The water is pressurized with a pump, distributed to areas of the fields through pipes or hoses, and sprayed across the soil surface with rotating nozzles or sprayers.

Figure 3-2: Surface irrigation

Types of sprinkler systems depend on the layout of the distribution pipelines and the way they are moved (i.e., solid set, hand move, center pivot, or rain gun). Micro irrigation systems, also called drip or trickle systems use small tubing to deliver water to individual plants or groups of plants. These systems use regularly spaced emitters on or in the tubing to drip or spray water onto or into the soil. Micro irrigation systems are categorized by the type of emitters (i.e., drip or micro spray).

3.1.3 Planning and selection of irrigation system

Most irrigation systems are installed to make a profit for the owner, thus it is necessary to determine the costs and factors that will ensure a profitable irrigation. The following are considered.

Physical resources: Consideration includes physical resources of:

a. *Water supply*: Location of water source available quantity, timely distribution, source of water supply, flow rate, quantity and reliability.
b. *Soil* e.g. structures, soil type, infiltration rate, texture drainage, erodibility factor etc.
c. *Topography*: Micro and macro relief features, slope
d. *Climate*: crop – system interaction
e. *Capital and labour*
f. *Energy supply* (pumping)

Risk conditions for irrigation

The followings are characteristics of soils that pose a high risk of environmental degradation when irrigated

1. Steep slope
2. Excessively slow or fast permeability
3. Shallow soils
4. Soils with a subsoil layer that restricts root growth
5. High water table
6. High saline or sodic soils

Annual operating and maintenance costs: This is the annual operation costs and maintenance: Cost of owing/renting, operating and maintaining the system and system components as well as cost of production of irrigated crop. Energy cost, water cost, taxes insurance, all land development costs e.g. land clearing conservation structures, capital investment on stores, farm houses, roads and drainage systems.

Depreciation based on expected life of the element (average of 2000 hours of use per year) longer life span is used for complex projects. An analysis period of 20, 25, 30 years are common for analysis of on -farm irrigation systems.

Other economic and financial feasibility values of farm project

Economic feasibility of a project assesses the economic viability of the entire project and assists in selecting a suitable system.

Financial feasibility: Financial feasibility of a project assesses the financial conditions that will be encountered in developing and operating the project systems on the farm.

This indicates the actual year by year costs and revenues expected after the development of the enterprises.

Factors influencing economic and financial feasibility of a particular project include size of farm (area in hectares), type of cropping system to be adopted, fund available etc.

Economic feasibility evaluation: This requires estimates of all costs and returns expected. This is determined by developing costs and returns as annual values on a common time basis.

Annual cost values: This is the depreciation of system components and cost of interest on capital investment (initial costs).

Present worth (Pwt): This is a term required to carry out annual cost calculation where individual items in the system are depreciable in less time than the adopted analysis period given by:

$$Pwt = (1 + i)^n \dots \dots \dots \dots \dots \dots \dots .3.7$$

Present worth value (PWV) is expressed by the following equations

$$Pwv = S(1 + i)^n = S(Pwt) \dots \dots \dots \dots \dots .3.8$$

Where
   S = Number of years the cost incurred in future
   I = Interest rate
   n = Replacement cost.

Amortization on capital invested on any system is determined from the present worth value of the investment plus the interest during analysis period.

$$Amortization = Pwv + i = S(1 + i)^n + i \dots \dots \dots \dots \dots .3.9$$

Variable costs: This include annual operating costs such as labour and other items involved in the operation of the system, maintenance and repair costs estimated as percentage of the initial investment cost in the various components of the system.

Cost escalation: Labour and/or energy costs may increase more rapidly than the cost of other inputs and this may affect initial design selection of a particular system component design and should be assessed by a sensitivity analysis using escalated annual variable costs.

Economic evaluation benefit: Benefit cost ratio is the annual benefits divided by annual costs. This ratio indicates the attractiveness of the enterprises.

Annual costs: loan repayment, variable costs (energy, labour taxes).

Annual revenue: Sales of farm produce.

Environmental and social considerations

Environmental assessment is necessary in planning and selection of irrigation system in order to identify changes in environment as a result of the system development. Environmental inventory include:

1. *Physical and chemical characteristics:* Earth (soil, land form) water (quality, recharged etc), atmosphere (climate), processes flooding, erosion, soil stabilities).
2. *Biological conditions:* Flora (trees, grasses, endangered species), fauna (animals – birds, fish etc).
3. *Cultural factors:* Kind use, recreation, aesthetics and human factors, cultural status (cultural patterns, health safety etc.).
4. *Ecological relationships:* Stalinization of soil, water etc.
5. *Other factors:* Other factors include site specific items.

Environmental problems encountered in project development include: water diversion, stream flow consideration, and salinization.

3.1.4 Types of irrigation systems

a. Surface irrigation

This methods include furrow, basin, border, contour ditches (wild flooding), and water spreading. Contour ditches (wild flooding), and water spreading basically consist of directing water diverted from ditches or watercourses onto sloping fields. They are primarily used to irrigate pasturelands and are generally very inefficient. The main surface methods are basin, furrow, and border irrigation.

Types of surface irrigation

Description of systems: The practice of surface irrigation is ancient and is used on more than 90% of the world's irrigated area. The sustainability of irrigated agriculture depends on improvements and innovations in surface irrigation methods, their appropriateness for the different systems, and their adoption infield practice.

i. Basin irrigation

Basin irrigation is the most commonly used system worldwide. It consists of applying water to leveled fields bounded by dikes. In flat areas, the basins are commonly rectangular and often have independent supply and drainage. When water saving irrigation is practiced in tropical areas where water ponding is not required for temperature regulation, water is applied only to keep the soil near saturation. In modern rice basins, laser leveling is used, basin size often exceeds 1 ha, and each basin has independent supply and drainage facilities. Rice transplanting may be replaced by mechanical seeding, sometimes in dry soil.

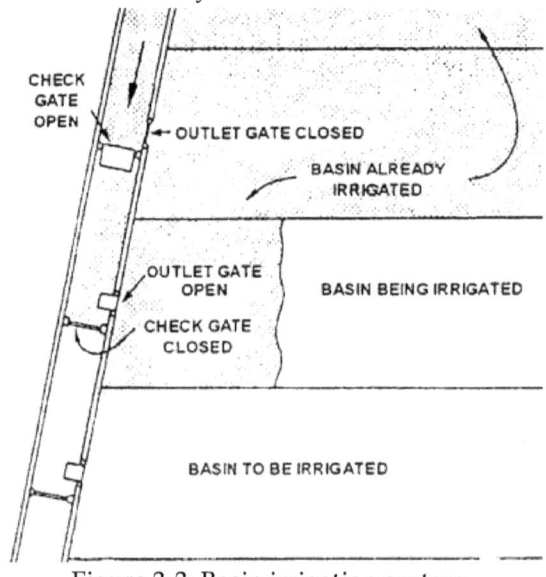

Figure 3-3: Basin irrigation system

Surface drainage often is not provided with basin irrigation. This simplifies the layout of the fields and the water delivery channels, but can result in water logging and soil aeration problems if soil infiltration is low and rainfall is high or irrigations are large. Where rainfall may be high during cropping, a network of surface drainage channels should be provided. Because there is normally no runoff from basins, quantifying irrigation applications requires only measurement of inflows and irrigation time. A desired net application can be preset if flow rate is known.

ii. Furrow irrigation

In furrow irrigation, small regular channels direct the water across the field (Figures 3-4 & 3-5). These channels, called furrows or corrugations, serve both to convey water across the field and as the surface through which infiltration occurs.

Figure 3-4: Irrigated vegetable field

Because conveyance and infiltration are two opposing purposes, designing and operating furrow systems always requires balancing the trade-off between quickly conveying the water across the field and maintaining the flow long enough to infiltrate adequate water.

Figure 3-5: Furrow irrigation with siphon tubes from a concrete-lined ditch

Furrow irrigation is used most commonly for row crops planted on beds or ridges. Furrows may be formed between each plant row (bed), or between alternate rows. Alternate furrows may be used for any given irrigation. Furrow spacing in broadcast and row crops varies from 0.6 to 1.6 m. Furrows in orchards may be up to 3 m apart. Small furrows or corrugations also are used in close-growing crops such as small grains, pastures, and forage.

iii. Border irrigation

In border irrigation, the field is divided into sloping strips of land separated by parallel border dikes or ridges. Water is applied at the upstream end of the strip and moves as a sheet down the border. The *border strip* is an area of land bounded by two border ridges, dikes, or ditches that guide the irrigation stream from a point or points of application to the ends of the strip.

The *border ditch system* uses a ditch as a divider between individual strips. The ditch carries irrigation water which is applied at different locations along the entire length of the field.

The *border dike system* uses dikes on both sides of the strip and the irrigation water is supplied only at the upper end of the strip.

Border irrigation is used primarily for close-growing crops, such as small grains, pastures, and fodder, and for orchards and vineyards. The method is best adapted to areas with low slopes (less than $0.005mm^{-1}$), soils with moderate infiltration rates, and large water supply rates. These conditions allow large borders that are practical to farm. Border width is determined by cross slope and available flow rates. The elevation difference across a border should be less than 30% of the flow depth to ensure adequate water coverage. Thus, border width is limited by field cross slope or by the amount of land movement required to eliminate cross slope.

Borders width typically varies from 5 to 60 m. Border length affects advance time and thus irrigation cutoff time. Longer borders require longer irrigation times and result in greater irrigation depths. Borders up to 400m long are used where infiltration rates are moderately low.

b. Overhead irrigation systems

Examples of this method of irrigation include the followings:

i. Sprinkler irrigation

Sprinkler irrigation is a method in which water is applied above plant foliage, high enough that water will be distributed freely and without obstruction. The sprinklers are connected at the top of riser equally spaced along the laterals. The laterals convey water under pressure head and divide it efficiently among the risers.

Figure 3-6: Overhead laterals and sprinklers lay top of the crop

The sprinkler distributed the water over a circular or rectangular area under light wind. The water jet coming out of the nozzles breaks up into drops which spread over the sprinkled area. At the nozzle, the pressure head is converted into velocity head giving the water jet its initial velocity.

The combination of pressure head and nozzle diameter determine the intensity of drop formation and the distribution over the wetted area. Sprinkler irrigation in agriculture began with the development of impact sprinklers and lightweight steel pipe with quick couplers. In the 1950s, improved sprinklers, aluminum pipe, and more efficient pumping plants reduced the cost and labour requirements and increased the usefulness of sprinkler irrigation.

In the 1960s, the development of moving systems, namely the center pivot, provided for moderate-cost and high-frequency irrigation. Additional sprinkler innovations are continually being introduced that reduce labour, increase the efficiency of sprinkling, and adapt the method to a wider range of soils, topographies, and crops. The main elements of a sprinkler system are:

There are many types of sprinkler systems, but all have the following basic components:

1. The *pump* draws water from the source, such as a reservoir, borehole, canal, or stream, and delivers it to the irrigation system at the required pressure. It is driven by an internal combustion engine or electric motor. If the water supply is pressurized, the pump may not be needed.
2. The *mainline* is a pipe that delivers water from the pump to the laterals. In some cases the mainline is placed below ground and is permanent. In others, portable mainline laid on the surface can be moved from field to field. Buried mainlines usually are made of coated steel, asbestos-cement, or plastic. Portable pipes

usually are made of lightweight aluminum alloy, galvanized steel or plastic. In large fields the mainline supplies one or more submains that deliver the water to the laterals.

Figure 3-7: Irrigation pipes

3. The *lateral* pipeline delivers water from the mainline to the sprinklers. It can be portable or permanent and may be made of materials similar to those of the mainline, but is usually smaller. In continuous-move systems, the lateral moves while irrigating.

Figure 3-8: Lateral pipeline

4. *Risers:* These are vertical pipeline installed on the laterals bearing the sprinkler head. The risers are connected to the laterals with couplings.

Figure 3-9: Sprinkler riser

5. *Sprinklers* spray the water across the soil surface with the objective of uniform coverage. Sprinklers can be adapted to most climatic conditions, but high wind conditions decrease distribution uniformity and increase evaporation losses, especially when combined with high temperatures and low air humidity. Although sprinkling is adaptable to most topographic conditions, large elevation differences result in non-uniform application unless pressure regulation devices are used.

Types of sprinkler heads and nozzles

Most sprinklers are designed to give a circular wetting pattern. Sprinkler heads are designed to throw water in a full circle, a half circle, or a quarter circle. The distance from the sprinkler to the outer edge of that circle is called the throw or wetted radius. Nozzle design, size, and pressure determine the pattern of wetted diameter. Sprinklers may have special features that allow them to irrigate only a part of the circle.

Traditionally, sprinkler heads are grouped into two broad categories based on the method they use to distribute the water, these are; the spray type sprinklers and rotor type sprinklers. However new technologies are blurring the traditional boundaries between the types.

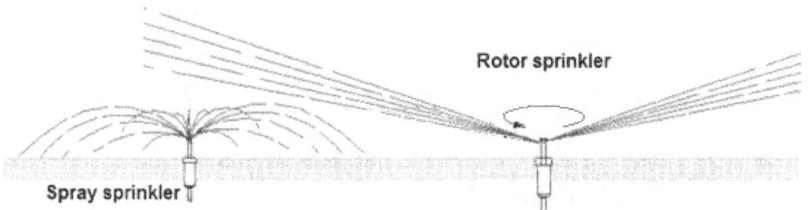

Figure 3-10: Types of sprinkler heads

Spray type sprinklers: More properly called "fixed spray heads," these are the small heads that spray a fan-shaped pattern of water. Think of a shower nozzle. Most use interchangeable nozzles installed on the sprinkler which determine the pattern (half circle, full circle, etc.) and the radius of the water throw. Some specialty patterns are available for long, narrow areas.

Recent adaptations of spray heads use rotating or wobbling plates with curved grooves that turn the plate by jet reaction. These variations are designed to increase throw at low pressures and improve water distribution patterns. Most manufacturers have classified the rotary nozzles, and rotators as types of spray sprinklers.

Figure 3-12: Spray patterns

Spray heads discharge a jet of water vertically from a nozzle onto an impingement plate that redirects it into a circular pattern. This plate may be smooth or serrated and have a flat, convex, or concave surface, depending on the desired pattern shape (throw and droplet sizes).

Spray head sprayers generally operate at low pressure and have smaller pattern diameters than impact sprinklers. They are commonly used on continuous-move lateral systems. Recent adaptations of spray heads use rotating or wobbling plates with curved grooves that turn the plate by jet reaction. These variations are designed to increase throw at low pressures and improve water distribution patterns.

Rotary or rotator nozzles sprinkler heads: These heads are called rotary nozzles because they are a very small rotor that is the same size as the standard nozzle on a spray-type sprinkler. Thus they fit onto the smaller, and less expensive, spray head pop-up bodies.

Rotary/rotator nozzles are more efficient than traditional spray heads because they produce less "mist" that evaporates before it reaches the ground. Thus they are often promoted for use in place of standard spray heads by water conservation agencies.

Figure 3-13: Rotary nozzles

These rotary nozzles have a radius generally between 15 and 35 feet. The exact distance depends on the model. They all use multiple streams of water that rotate around the nozzle and look like rotating spider legs. Rotary sprinkler generally operate at low pressure and have smaller pattern diameters than impact sprinklers. They are commonly used on continuous-move lateral systems.

Rotor type sprinkler heads: Rotor is the term used to describe the various sprinklers which operate by rotating streams of water back and forth or in circles over the landscape. The common example is the "impact" rotor sprinkler which moves back and forth firing bursts of water.

i. *Impact sprinkler* rotation is caused by the water jet impinging on a spring-loaded swing arm (Figure 3-14). The water jet impulse forces the spring arm sideways. The spring returns the arm, which impacts the body of the sprinkler, rotating it a few degrees between 22° and 28°. Then the cycle repeats. The jet breaks up into small drops as it travels through the air, and falls to the ground like natural rainfall. The rotational speed (1 to 3 rotations per minute) is controlled by the swing-arm weight and spring tension.

Figure 3-14: Impact sprinkler heads

Good water distribution is dependent on maintaining water pressure (jet velocity) within the range that produces the proper droplet sizes. The water jet commonly discharges at an angle above horizontal

ii. *Gear driven rotary sprinklers:* The impact rotors are rapidly being replaced now by gear driven rotors which are very quiet, lower maintenance, and much smaller in size. The pressurized water entering the sprinkler rotates a small water turbine which, through reducing gears, provides for slow, continuous sprinkler rotation. These new turbine and gear driven rotors have one or more streams of water which move silently across the landscape. Gear-drive mechanisms require clean water to prevent clogging and wear.

Figure 3-15: Gear driven rotary head

iii. *Stream rotors*: This is a type of rotor sprinkler having multiple streams of water that rotate around the rotor, one following the other. The streams of water always rotate in the same direction. The result is an interesting appearance, often described as spider legs moving around in a circle.

Figure 3-16: Stream rotor head

Stream rotors are popular for use on steep slopes. This is because the small streams apply water more slowly to the ground than either spray heads or standard rotors. A slower application rate of the water allows more time for it to soak into the ground, which results in less water run-off and related soil erosion.

Classification of sprinkler system

Sprinklers irrigation systems can be divided broadly into set and continuous-move systems. In *set systems*, the sprinklers remain at a fixed position while irrigating; in *continuous-move systems*, the sprinklers operate while the lateral is moving in either a circular or a straight path. Set systems include *solid set or permanent systems* as well as *periodic-move systems*, which are moved between irrigations, such as hand-move and wheel-line laterals and hose-fed sprinklers. The principal continuous-move systems are center-pivot and linear-move laterals, and traveling rain gun sprinklers.

Fixed (permanent) systems

When sufficient laterals and sprinklers are provided to cover the whole irrigated area so that no equipment needs to be moved, the system is called a *solid-set* system. For annual crops, the portable pipes and sprinklers are laid out after planting and remain in the field throughout the irrigation season. The equipment is removed from the field before harvesting. In perennial crops such as orchards, laterals and sprinklers often are left in place from season to season. Permanent systems often are buried below ground but they also may be in case of over tree irrigation for chemigation.

Semi permanent systems

Sprinkler systems have been developed with the advantages of both portable and fixed equipment to combine both low capital costs and low labor requirements. These often are referred to as semi permanent systems and the most commonly used are the pipe-grid and hose-pull systems. These systems are designed to reduce the number or size of laterals.

Pipe-grid systems

These are similar in many aspects to fixed systems. Small-diameter laterals (about 25 mm) are used to keep system costs low. Laterals are laid out over the whole field and they remain in place throughout the irrigation season. In general, two sprinklers are connected to each lateral, one near the end, the other near the middle. When the irrigation depth has been applied, each sprinkler is disconnected and moved along the lateral to the next position. This procedure is repeated until the whole field has been irrigated. A typical system would involve at least two sprinkler moves on every lateral each day.

Hose-pull systems

Originally developed for orchard under tree irrigation, these systems now are being used for some row crops (Figure 3-17). The mainline and laterals usually are permanently installed, either on or below the ground surface, but also can be portable.

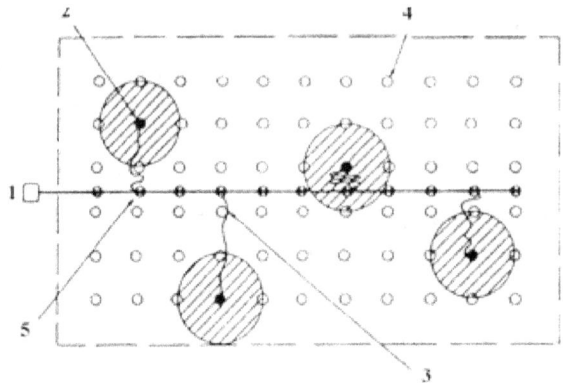

Figure 3-17: Hose-pull system

[(1) Pump, 2) sprinkler, 3) flexible hose, 4) sprinkler positions, 5) lateral]

Small-diameter plastic hoses supply water from the lateral to one or two sprinklers. The hose length is normally restricted to about 50 m because of friction losses. Initially, the sprinkler is placed in the farthest position and remains there until the irrigation depth is applied. Then it is pulled along to the next position and so on until irrigation is complete.

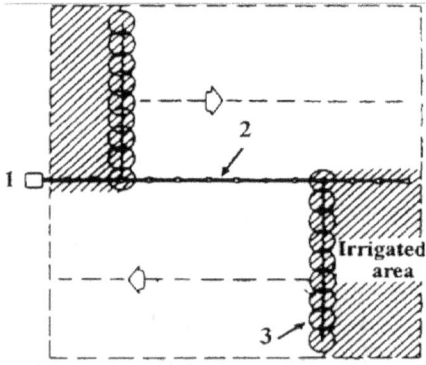

Figure 3-18: Hand-moved system

[1) pump, 2) mainline, 3) lateral]

Portable systems

i. Hand-moved system

These are designed to be moved by hand. The lateral is usually in aluminum or plastic pipe between 50mm and 100mm in diameter and 9 to 12m long, so that it can be moved easily by one person.

The laterals remain in position until irrigation is complete. The pump then is stopped and the lateral disconnected from the mainline, drained, dismantled, moved by hand to the next point on the mainline, and reassembled. Usually, the lateral is moved between one and four times each day. It gradually is moved around the field until the whole field is irrigated.

ii. Center-pivot systems

This consists of single galvanized steel lateral, one end anchored to fixed pivot structure and the other end continuously moving around the pivot while applying water. Lateral pipe diameters range from 100 to 250 mm. The lateral is supported using cables or trusses as much as 3 m above the ground on A-shaped steel frames mounted on wheels (Figure 3-19). The frames are spaced approximately 30 m apart. Laterals vary in length from 100 to 800 m. A common lateral length is 400 m, which irrigates up to 50 ha.

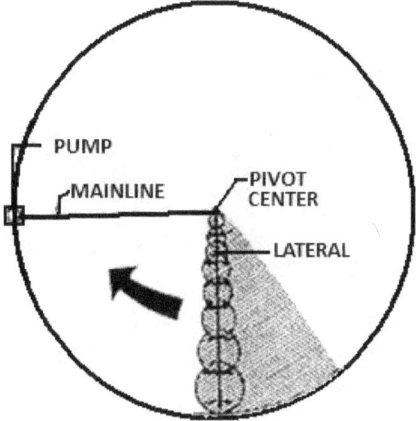

Figure 3-19: Center-pivot system

Water is supplied to the center pivot by a buried mainline or directly from a well located near the pivot point. Water flows through a swivel joint to the rotating lateral and sprinklers .When irrigating, the lateral rotates continuously about the pivot, wetting a circular area. The rate of laterals rotation about the pivot determines the rate of water application. , The slower the speed laterals rotation, the more water that is applied to the field. Typical applied depths vary from 5 to 30 mm. A center pivot lateral is therefore a system that can effectively apply light, frequent irrigations. The drive unit can be powered by:

- Hydraulic water-driven piston – driver system, rotating sprinkler and turbine drive systems, or others (chain and sprocket mechanisms).
- Electric-drive system
- Oil powered system
- Cable drive system

Water application rate in pivot system: rate is affected by sprinkler type, spacing along lateral and diameter of area covered from individual sprinkler. Application rate is low near the pivot and increase progressively toward outer end because length of time it takes water to be applied pen unit length of lateral decreases from pinot to enter end.

iii. Microspray irrigation

Microspray was developed to wet a larger percentage of the rooting area of tree crops than was practical with drip irrigation. Microspray heads are small versions of low-pressure sprinkler heads. In the most common type, a small, vertical water jet from a small orifice nozzle impacts a deflector plate that diverts the jet into a horizontal pattern. *The nozzle size* determines the flow rate (at a given pressure) and the *deflector-plate* shape determines the *spray pattern*. Microspray emitters spray water over 2 to 6 m diameter circles or partial circles.

Figure 3-20: Microspray irrigation showing a half-circle pattern sprayer and tubing

Microsprayers usually are inserted into the end of short pieces of 6 mm tubing and held upright on a stake (Figure 3-20) or hung from suspended tubing. The small tubing is attached to the drip tubing laterals with small plastic barbed connectors. Microsprays are commonly used for widely spaced tree crops. They also may be used in greenhouses.

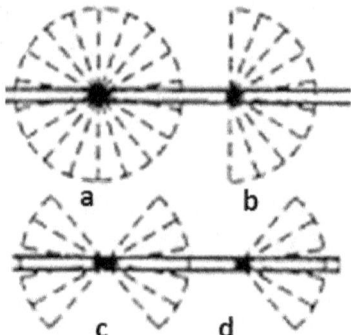

Figure 3-21: Typical microspray wetting patterns

In tree crops, one or two sprayers are used on each tree. The pattern usually is oriented so that the water does not spray on or at the base of the trunk, to reduce disease problems (Figure 3-21).

Advantages of microirrigation systems

Advantages of microirrigation systems include:

i. Potential to reduce irrigation water use and corresponding operating costs because water can be applied only where the crop roots develop.
ii. It has been shown to increase the yield and quality of some crops. This is most likely the result of maintaining near-optimum water and fertility conditions in the root zone.
iii. It can reduce the cost of labour because the systems need only to be maintained and managed, not tended. Operation usually is accomplished by automatic timing devices, but the emitters and system controls should be inspected frequently.
iv. A greater control over fertilizer placement and timing through fertigation with microirrigation improves fertilizer efficiency and reduces pollution hazards associated with fertilizers.
v. It can reduce weed growth and the incidence of some diseases because foliage and much of the soil surface are not wetted. This reduces costs of labor and chemicals to control weeds and diseases and reduces related pollution hazards.
vi. Well-designed microirrigation systems can operate efficiently on almost any topography.
vii. Problem soils with low infiltration rates, low water-holding capacity, and variable textures and profiles can be irrigated efficiently.
viii. It usually requires lower operating pressure and thus less energy than sprinkler systems.

Disadvantages of microirrigation systems

Microirrigation has the following disadvantages:

i. Equipment costs usually are higher than for surface irrigation systems and may be higher than for sprinkler systems.
ii. Equipment often is complex and requires frequent monitoring to ensure good performance.
iii. Energy costs to pressurize the system are higher than with surface irrigation.

iv. Because emitter outlets are very small, they can become clogged by particles of mineral or organic matter. Clogging reduces discharge rates and the water distribution uniformity; thus filtration is required in most cases. Iron oxide, calcium carbonate, algae, and microbial slimes may be problems requiring chemical treatment of the water to prevent clogging.

v. Because microirrigation systems operate at low pressures, varying field topography can result in significant pressure variations and irrigation non-uniformity. Careful design and pressure regulation are required on undulating land.

vi. Some crops do not germinate well with drip irrigation and especially with subsurface drip. A second method of irrigation, usually portable sprinklers, may be required for germination.

vii. Salts may concentrate at the soil surface and between emitters and become a potential hazard. Localized salt accumulation can hinder crop germination. Light rains can leach accumulated salts downward into the root zone. Irrigation should continue on schedule unless adequate rain has fallen to ensure leaching of salts below the root zone.

viii. Salts also concentrate below the surface at the perimeter of the wetted bulbs. Too much drying of the soil between irrigations may allow the movement of water and salts back toward the inner bulb. To avoid this damage, irrigation must be frequent under saline conditions.

ix. If unexpected events (equipment failure, power failure, or water-supply interruption) interrupt irrigation, crop damage may occur quickly because roots use only a small volume of wetted soil. At least 33% of the total potential root zone should be wetted. Careful system maintenance and a secure water supply are a must.

x. When a main supply line breaks or the filtration system malfunctions, contaminants may enter the system, resulting in emitter clogging. Secondary filters can be used to protect against these problems.

xi. Rodents and other small animals sometimes chew and damage polyethylene tubing. In some cases, animal damage disallows laying tubing or emitters on the soil surface. Burrowing rodents also may damage subsurface tubing. Rodent control may be necessary to reduce the problem.

Advantages of sprinkler irrigation

Sprinkler irrigation has the following advantages compared to surface irrigation:

i. Properly designed and operated sprinkler irrigation systems can give high seasonal irrigation efficiencies and save water.

ii. Sprinkler irrigation performance is not dependent on soil infiltration (as long as application rate does not exceed infiltration rate), and thus is dependable and predictable.

iii. Soils with variable textures and profiles can be efficiently irrigated.

iv. Land leveling is not required; shallow soils that cannot be graded for surface irrigation without detrimental results can be irrigated.

v. Steep and rolling topography can be irrigated without producing runoff or erosion.

vi. Light, frequent irrigations, such as for germination of a crop, can be given.

vii. Sprinkler systems can effectively use small, continuous streams of water, such as from springs and small-tube or dug wells.

viii. Periodic-move sprinkler systems require only unskilled labor; irrigation management decisions are made by the manager.

ix. Fixed sprinkler systems can be used to control weather extremes by increasing air humidity, cooling the crop, and reducing freeze damage.

x. Cultural practices such as conservation tillage and residue management can be used easily under sprinkler irrigation.

Limitations of sprinkler irrigation

Sprinkler irrigation has the following limitations:

i. Initial costs are higher than for surface irrigation systems unless extensive land grading costs are required.

ii. Energy costs for pressurizing water are a significant expense, depending on the pressure requirements of sprinklers used and power costs.

iii. When water is not continuously available at a sufficient, constant rate, the use of a storage reservoir is required.

iv. Windy and dry conditions cause water loss by evaporation and wind drift.

v. Irregular field shapes are more expensive and less convenient, especially for mechanized sprinkler systems.

vi. Certain waters are corrosive to the metal pipes used in mainline and laterals.

vii. Water containing trash or sand must be cleaned to avoid clogging and nozzle wear.

viii. Sprinkler irrigation water containing salts may cause problems because salts drying on the leaves affect some crops. High concentrations of bicarbonates in irrigation water may affect the quality of fruits. Sodium or chloride concentration in the irrigation water exceeding 70 or 105 parts per million (ppm), respectively, may injure some fruit crops.

ix.   The high humidity and wet foliage created by sprinkling is conducive to some fungal and mold diseases.

b. *Subsurface irrigation system*

In unique situations water may be applied below surface of the soil. There are two types of sub surface irrigation.

i.   Water table maintenance method of sub-irrigation maintains a water table allowing the water to move up to the root zones by capillary action. It is applicable where the soil in the plant root zone is permeable and the rest is either a continuous impermeable layer or natural water table below the root zone.

ii. The second method of sub-irrigation introduces water into the soil through perforated pipes. Plastic pipes "are laid 4 – 6" (10 – 12.5cm) deep in gravel envelopes and on 30" (75cm) centers. Water introduced into these pipes move throughout the root zone by capillary action.

Subsurface drip (trickle) irrigation (SDI)

Trickle (drip) irrigation is the latest type of water supply to plant in the technology market. Water is applied through narrow plastic tubes stuck in the side of the plant being irrigated. Water is applied directly to the base of the plant and there is little wastage as happens through other methods. Water application may be cut be 50% using this method.

Cost of pumping and distributing water is minimal. This system is still being used in small pilot projects because of the high cost of plastic piping.

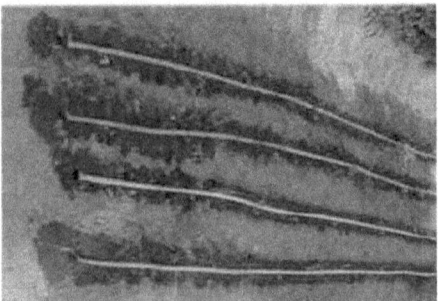

Figure 3-22: Wetting pattern of drip tape

Drip irrigation applies small amounts of water frequently to the soil area surrounding plant roots through flexible tubing with built in or attached emitters. Subsurface drip irrigation (SDI) delivers water underground directly to roots.

Advantages of SDI

Since water is applied directly to individual plant roots, SDI

1. Minimizes or eliminates evaporation,
2. Provides a uniform application of water to all crop plants,
3. Applies chemicals more efficiently and.
4. Reduces plant stress and increases crop yield. A carefully managed amount of water is applied, thereby avoiding deep percolation and runoff, while reducing salt accumulation.

Disadvantages of SDI

Although drip systems are very efficient, they do have some drawbacks.

1. Because they may clog and are susceptible to damage by rodents, insects, and sedimentation, they must be checked regularly.
2. A good filtration system is essential for proper performance of a drip system.
3. Hard water should be treated to discourage mineral build-up.
4. New systems are expensive, and must be designed to suit crops and local soil and climate conditions.
5. A reliable, continuous water supply is necessary to run a drip system, and
6. Proper irrigation management and furrow shaping is necessary to prevent salt buildup.

Components of drip irrigation system

1. A *"head"* connected to the main water supply to the field which includes filters, valves, couplings, water meter, pressure quake conducting pipes
2. Conducting pipes

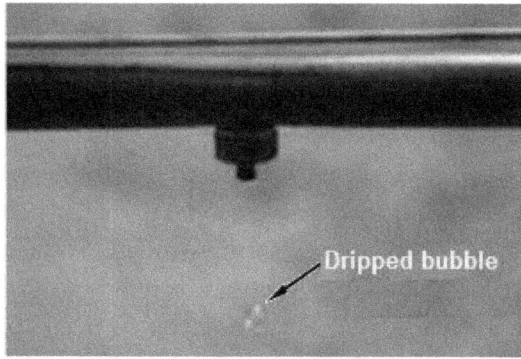

Figure 3-23: Drip nipple

3. *Nozzles* with small perforations and an arrangement to reduce pressure so that the water leaves the system in the form of drops.
4. *Distribution tubes* (branches) of small dimmers ($\frac{1}{2}$ - $2/3$ inches) connected to the conducting pipes.
5. A fertilizer apparatus connected to head through which $1/3$ to $\frac{1}{4}$ of the total water flow passes.

Low energy precision application (LEPA)

LEPA irrigation systems distribute water directly to the furrow at very low pressure (6-10psi) through sprinklers positioned 12-18 inches above ground level. Conventional high pressure impact sprinklers are positioned 5-7 ft. above the ground, so they are very susceptible to spray evaporation and to wind-drift, causing high water loss and uneven water distribution.

LEPA systems apply water in streams rather than fine mists to eliminate wind-drift and to reduce spray evaporation, deep percolation and under watering. LEPA irrigation systems further reduce evaporation by applying water in bubble patterns, or by using drag hoses or drag socks to deliver water directly to the furrow.

Advantages of LEPA systems

1. They concentrate water on a smaller area and increase the water application rate on the area covered. Therefore, the application rate must be monitored closely to follow the soil intake curve, and furrow diking should be used to prevent runoff. In addition to water savings,
2. They use much less energy (at least 30% less than conventional systems), which reduces fuel consumption and operating costs.
3. They reduced disease problems due to less wetting of foliage, and easier application of chemicals. Both lateral move (side roll) and center pivot systems can be readily converted to LEPA irrigation. Variable flow nozzles adjust flow from a computer to match microclimate conditions.

Management practices

Correct management of a LEPA system is essential to realize potential water savings. Farmers who replace older irrigation systems with LEPA sprinklers should adjust their management practices so that they do not continue to use excess water. If the pivot system does not have a digital control box showing the amount of water applied, meters should be installed or readings from portable meters should be requested from the local water district to accurately determine how much water is being applied.

3.1.5 Design of irrigation systems

System selection

The first decision that an irrigation designer must make is the selection of the irrigation method. This choice depends on both physical and socioeconomic factors, including the cost, availability, and quality of the water supply; the soil type; the field topography and geometry; the crop type and value; the labour cost and availability, material costs, energy costs; and the practicability and availability of the various technologies (Keller and Bliesner., 1990).

Surface irrigation

Based on technical feasibility and economics, surface system is the cheapest. The followings are recognized.

a. *Flooding irrigation:* wild flooding of field by pool of water during heavy rains.
b. *Basin irrigation-* field is divided into small units (basins) so that each has a nearly leaves surface. Small banks of earth 30-50cm high called levies bund are consternates with inlet and out let controls for water. Most of the rice in world had been grown in basins known as in the east as paddies. Many other elopes such as cotton grains vegetables are best surfed.
c. *Furrow irrigation:* small channels (furrows) carry water down or across the slope of the land wet the soil. This method is best suited to deep moderately permeable soils with uniform slope relatively to crops which are cultivated in rows such as vegetables. Tomatoes, cotton, potatoes and maize.
d. *Border irrigation:* suitable for large field units of 4.0 hectares or more. A field is divided into series of trigs 3-30m wide, 100-800m long by borders or low banks with an even, moderate slope along the length. The levies forming the feels to the slips should be 20-25cm high after settlement and triangular in cross section. Water is supplied from channels through turn-outs (wooden or concrete). Gated pipes placed on the channel bank with gate control or portable siphons placed over the bank.

Design considerations of field layout for surface systems

1.   Identify canals (main and secondary) and field ditches on project site

2. Determines crop water requirement and hence size of supply and field canals and ditches
3. In designing the irrigation stream, take note of the followings

    i. Irrigation principle of surface system during imitable advance stream, flooring down the field
    ii. The welting period when whole length of run is infiltrating water
    iii. The recession stream after water is cut off. Normal practices is to introduce large stream initially to reach and of field quickly and them cut back supply to such a flow that will be maintained until soil is wetted to the required depth.

4. Design operation to aim at equalizing the opportunity time for water intake at all point of the run.

Design of furrow irrigation layout

1. *Shape and spacing of furrow*: ridge height and slope determine carrying capacity of furrows. Ridge height 0.15-0.4m furrow spacing based on crop spacing modified if necessary to obtain adequate lateral wetting (0.3-1.8m). shape depends on shape (between 2-65% along the furrow is advisable)
2. *Advanced furrow stream*: maximum non-erosive flow untying furrow capacity to ensure early wetting of furrow is advised. Empirical relation 0.6/8 lts / second (8 is percolates shape) is used.

Application depth, d is given by

$$d = \frac{(360 \times g \times t)}{w \times l} \dots \dots \dots \dots .3.10$$

Where
    g = stream size (1/s)
    t = duration (hrs),
    w = furrow spacing (m)
    L = furrow length

3. *Length of furrow*: this is a function of soil. In light soils, 60m I s recommended and 200m for heavy soil.

Design of border strip irrigation

Strip width: factors that influence width includes: elimination of cross slope, appropriate stream, and implication of mechanization. Recommended rates are given in literatures

Selection of irrigation stream: maximum non-erosive advance stream is suggested. Suggested values should serve as a guide for on field test run applied to non-sod forming crops such as pasture and alfalfa, they, may be doubled.

Strip length by field test: chosen design stream is used ad cut backs were applied at 2/3 or ¾ strip length to achieve the desired aim of no run off and uniform spread across strip.

Design consideration of basin irrigation

1. *Deep percolation loss: this* is not a serious problem since irrigation is carried out on impermeable soils.
2. *Size of basin:* Limited by crop as well as amount of grading necessary to obtain zero slopes in all direction.
3. *Time requirement:* Time required for water to cover the basin should be limited to one fourth the required intakes time for the not irrigation depth.
4. Determine the optimum relationship between stream size and the size of the check basin for the given soil type.

Sprinkler systems design

There are two stages involved in the design of sprinkler system thus: data collection and determination of system design factors.

1. Collection of agricultural and technical data
   a. Agricultural data include

      i. Type of crop
      ii. Water absorption pattern
      iii. Life span of the crops
      iv. Water/moisture sensitive stages of the crop

   b. Technical data include

      i. Precipitation – rainfall distribution pattern
      ii. Humidity
      iii. Temperature

iv. Wind speed and direction

2. Determination of sprinkler system design (data) factors: These include:

    a. Topographic map of area to be irrigated. The map should include elevations, existing roads, fence, existing pipelines, water sources etc.
    b. Water source: Such information below is required

        i. Type of water source; artificial reservoir, lake, pond, canal, river, underground waters etc.
        ii. Water source discharge and recharge capacity.
        iii. Water quality – pressure of foreign particles and other trace element that could course corrosion in pipes. This often have effect on Hazen – William
        iv. Factors required in design

3. Miscellaneous

    a. Available equipment
    b. System efficiency
    c. Manpower and level
    d. Sprinkler lateral design
    e. Main pipelines design and pump solution

4. Other factors to be considered in design include:

    a. *Soil*: Soil properties to be defined include; soil profile, soil depth, existence of hard pan etc.
        i. Soil texture, structure, aeration etc.
        ii. Field capacity, wilting point, bulk density etc.
        iii. Soil infiltration capacity.
    b. *System layout*: Pump location and the choice of system (permanent, semi-portable or portable) etc.
    c. *Selection of sprinkler and nozzle*: This is done on the basis of discharge, pressure, spacing, type and performance of sprinkler.

    The spray must not be too fine otherwise it will be carried away as drift, it must not be two heavy otherwise it may cause erosion and damage crops tender canopy. Specifications of spray pattern and pressure are given in manufactures guide.

Steps in design

1 *Determination of gross irrigation requirement and the effective irrigation interval*

   a. Find the readily available moisture content on volume basis.
   b. Determine the net moisture requirement.

$$WR = mc_{av} \; x \; \rho \; x \; mc_{ct} \; ....3.11$$

Where
   WR= Net requirement
   mc_{av} = Available moisture
   ρ = Bulk density
   mc_{cr} = Critical moisture content

   c. Determine gross water requirement

$$G_{IR} = \frac{N_{IR}}{\eta} \; ....3.12$$

Where
   G_{IR} = Gross irrigation requirement
   N_{IR} = Net irrigation requirement
   η = Field efficiency of the system

   d. Calculate Irrigation interval D (days)

$$D = \frac{WR}{CU} \; ....3.12$$

Where
   CU = Consumptive use

2 *Lateral line requirement*: The number of lateral required depends on how big the farm is and what the irrigating time per set is. The number of lateral operating is affected by the total amount of water available.

3 *Sprinkler operating time and number of shifts per day*: This depends on the application rate of the sprinkler and the total amount of water to be applied. For instance, if you want to apply 80mm of water to irrigate night and day, and for an application you rate of 8mm/hr, then you need 10hrs per application. You can have 2 shifts and 4hrs left for other operations like lifting of laterals.

4   *Design of the laterals*: In the design of laterals take the following steps:

    a. Determine the length of lateral
    b. Know the spacing of the sprinkler on the lateral and between the lateral.
    c. Determine the number of sprinklers along the lateral and the discharge per sprinkler.
    d. Determine the pressure head.
    e. Determine the allowable head loss along the lateral
    f. Determine the C and K values for the material for lateral.

With the above information, select the best lateral diameter 2", 3" or 4". Allowable head loss T_c is based on rule of thumb practice of 20% of the operating pressure. Calculate the head loss in the diameter selected using appropriate head loss coefficients given by Hazen-William J value for each C and multiply by K factor.

5. *Design of submain:* For each field, the arrangement of laterals operating simultaneously along the submain is planned. One or more extreme situations depending on topography are then checked during which the maximum of laterals are operating. The design discharge load is set at maximum during this period. Taking frictional losses and other losses into consideration, the main pipe is designed and the pump capacity determined.

Sprinkler selection and spacing

After general system selection, the particular combination of nozzles, operating pressure, spacing of sprinklers along laterals, spacing of laterals along main (lateral shift distance).

i. *Net depth* d_n *(mm) and irrigation internal D (days)* are computed from depth of readily available moisture potentially stored in the root zone and the rate of crop waters use (peak crop consumption use).
ii. *Gross depth of application*

$$d_{gross} = \frac{d_n}{\eta_{app}} \; \ldots \ldots \ldots \ldots \ldots \ldots \ldots 3.13$$

    Where
    d_{gross} = Gross application depth
    d_n = Net Irrigation requirement
    η_{app} = Efficiency of application

Efficiency of application of 0.7 and 0.9 were assumed in hot and cold humid climates respectively. Minimum time T for the application of d_{gross} is given by

$$T = \frac{d_n}{\eta_{app}p} \ldots\ldots\ldots\ldots\ldots\ldots\ldots\ldots 3.14$$

Where

p = application rate

iii. The required discharge of individual sprinkler is a function of water application rate and the two way spacing of sprinkler may be determined by

$$Q = \frac{S_1 \times S_m \times I}{360} \ldots\ldots\ldots\ldots\ldots\ldots\ldots\ldots 3.15$$

Where

Q = Required discharge of individual sprinkler L/s
S_1 = Sprinkler spacing along lateral (m)
S_m = Lateral spacing along main (m)
I   = Optimum application rate cm/hr

(Manufacturers' sprinkler charts offer a choice of solution of the above equation.)

iv. Height of risers depends on maximum crop heights.
v. Sprinkler system capacity

$$Q = \frac{2780 \times A \times d_{gross}}{\eta_{app} \times H \times F} \ldots\ldots\ldots\ldots\ldots\ldots\ldots 3.16$$

Where
Q = Discharge capacity of pump L/S
A = Area of irrigated field (hectares)
d_{gross} = Gross depth of water application (cm)
η_{app} = Water application efficiency
H   = Number of operations in a day (hrs).

Design pressure head in the sprinkler and the main

Average head, H_a for design in a sprinkler line is

$$H_a = H_o + 1/4\, H_f \ .... 3.17$$

Where

H_a = Average frictional loss in lateral

H_f = Lateral frictional loss (within 20% of H_e)

H_o = Pressure at the sprinkler on the farthest end of line.

Pressure head, H_n at the main (lateral on level land) is given by

$$H_n = H_e + H_f = H_a + 3/4 H_f \ .... 3.18$$

Making allowance for elevation differences along lateral

$$H_n = h_a + 3/4 H_f \pm 3/4 H_e + H_r \ .... 3.19$$

Where

H_r = Riser height (m)

H_e = Maximum difference in elevation between first and last sprinkler (m). H_e is positive if lateral runs up slope and negative it runs down slope.

Center pivot sprinkling system design

Overall design procedures proposed by Dillon *at al.*, 1972 were stated below

1. Determine the radius of coverage of system (divide shortest dimension of field by two)
2. Determine peak water use and irrigation efficiency
3. Water flow required at the center pivot
4. Determine minimum time refigured to make one lateral revolution
5. Determine net depth of water applied to the root zone per revolution of the lateral.
6. Determine minimum design speed of travel of center- pivot lateral at which polentas run off starts.
7. Calculate the maximum depth of water that can be applied to root zone per revolution.

Quantities measured in center-pivot design include the followings

1. Irrigated area in the center pivot system is evaluated by the equation

$$A = \frac{\pi L_1 P_1}{K} \ (acre) \ \text{...\ ...\ ...\ ...\ ...\ ...\ ...\ ...\ ...} 3.20$$

Where

  L_1 = Effective irrigated radius of center pivot laterals (M)
  P_1 = The percent of full circle expresses in decimal.
  K = Constant 10,000 for metric unit and 43560 for English units.

2. Water needed at pivot to meet peak water use.

$$Q = KE_t \ x \ \frac{A \ x \ t}{E_i \ t_i}. \ \text{...\ ...\ ...\ ...\ ...\ ...\ ...\ ...} 3.21$$

Where

  Q = quantity of water ($l/2$) (gal/min)
  E_t = peak water use rate, mm/day (inches/ day)
  A = irrigated area, ha (acre)
  t = time between irrigation, (days)
  E_i = water application efficiency (70-80%)
  t_i = lateral operating time for one irrigation, (days)
  K = constant (0.116 metric unit, 18.9 English units)

3. Maximum application rate (elliptical patter)

$$P = K \ x \frac{Q}{L_g \ P_i} \ \text{...\ ...\ ...\ ...\ ...\ ...\ ...\ ...} 3.21$$

Where

  P = Maximum application rate of last few sprinklers, mm/hr (in/ hr)
  Q = Flow of water into center parrot, l/s (gal/min)
  K = Constant (4584 metric unit, 122.5 English unit)
  L_g = Effective irrigated radius of the center pivot lateral m (H)
  L_i = Radius of wetted area at the outer end of lateral.

4. Average gross depth of water applied by center pivot

$$D = K \cdot \frac{Q \cdot t_i}{A} \quad \text{.......... ... ...} 3.22$$

Where

  D = gross depth of water applied mm (ins)

  Q = flow of water into center pivot L/S (gal/min)

  A = Area irrigated, ha (acre)

  t_1 = Lateral operating time for one irrigation (hrs)

  k = Constant (0.36 metric, 0.00221 English unit).

5. Time to complete one lateral revolution

$$t_v = \frac{2\pi \ x \ L_3}{V} \quad \text{... ... ...} 3.23$$

Where

  L_3 = distance of end drive unit from pivot, m (ft)

  V = speed of travel of end drive unit m/hr (ft/hr).

3.2   Excess water removal (drainage)

Drainage is the removal of excess water from the land. In removing excess water in humid area, it is usually necessary to use either surface ditches, tiled drains or a combination of the two. Wetland is usually flat, has high fertility and does not have serious erosion problem. Drainage in humid areas often precedes land development while in arid region, it normally accompanies irrigation.

Figure 3-24: A rectangular drain

Land drainage deals with the control of water logging and soil salinization in agricultural lands. In flatlands, a first problem emerges if soil infiltration rates are low and rainfall or irrigation water stands on the ground surface in small depressions or at the edges of the irrigation basin.

This problem can be solved by leveling and smoothing the land and providing it with a uniform slope for excess water to flow through furrows or shallow ditches toward the surface drainage outlet. Surface water is discharged into a collector drain through pipes to prevent the erosion of the open ditch bank.

Land drainage has contributed to agricultural development in rain fed areas of the temperate regions, in irrigated lands of the arid and semiarid regions, and in the humid tropics.

In the temperate humid regions, land drainage promotes good aeration of the root zone and provides moisture appropriate in the topsoil to ensure workability. Therefore, in these regions, drainage has been an efficient means to increase crop production and to decrease farming costs.

In irrigated lands, drainage is indispensable to prevent the permanent hazard of water logging and salinization: may have adverse environmental effects such as conversion of wetlands to agricultural lands, deterioration of water quality, and landscape destruction.

Therefore, new drainage systems must be designed, constructed, and managed, taking into accounts not only agricultural objectives but environmental factors as well.

The principal purpose of drainage in irrigation

The principal purpose of drainage in irrigated region is to replace saline and alkaline soils by leaching and to prevent salinity problems by maintaining a low water table. Where salinity problems exist, land should not be developed for irrigation unless drainage facilities can be provided.

3.2.1 Stage investigations in drainage systems

Stage 1: The first phase of a drainage project is the identification of the waterlogged or salt affected lands and their further characterization for planning the reclamation procedure.

Stage 2: This phase is followed by design of the drainage system. During these two phases, some drainage investigations must be carried out by means of two fundamental studies: a soil survey and a hydrologic study, both based on a sound topographic survey. The climatic, soil, and hydrological data can be compiled in geographic information systems (GIS), by means of thematic maps, which can be continuously updated.

Stage 3: Additional investigations should be carried out to assess the environmental impact of the drainage project and to follow up its performance in order to address the protective measures that may be necessary. Investigations are also necessary to deal with the socioeconomic issues related to the drainage project.

Subsurface drainage systems

Subsurface drainage systems are installed in flatlands to control the groundwater level in order to achieve water and salt balances favorable for crop growth. The groundwater level can be controlled with a system of horizontal parallel drains (known as horizontal drainage) or by pumping groundwater in wells that penetrate into the aquifer (called vertical drainage). A horizontal subsurface drainage system consists of a network of parallel drains.

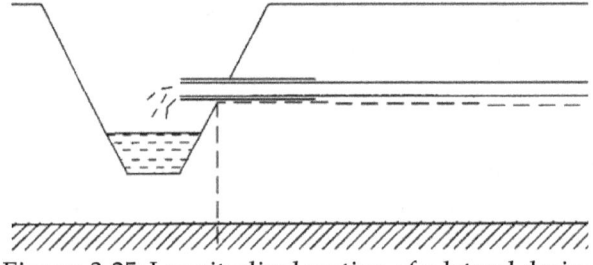

Figure: 3-25: Longitudinal section of a lateral drains

The groundwater flows toward the field lateral drains, which discharge into collector drains through an outlet pipe (Figure 3-26). The drainage water is conveyed through the network of collector drains and main canals toward the drainage outlet, where water is discharged into a water body by gravity through a tidal gate or by pumping.

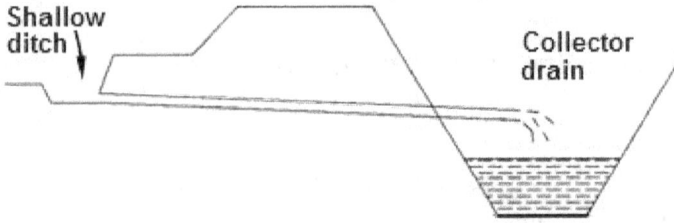

Figure 3-26: Outlet of surface runoff from a shallow ditch to a collector drain

Surface drainage systems

In flatlands, a drainage system is required to discharge standing water, caused by excessive rainfall or irrigation, over the soil surface. Two major components are essential: a graded and smoothed ground surface free of small depressions with an appropriate slope to enable the surface runoff to flow without producing soil erosion; and surface drains and shallow ditches to convey the drainage water toward the outlet into the main drainage system.

In sloping and undulating lands, the main issue is to prevent soil erosion due to overland flow; this can be achieved by land grading. In this way, surface runoff decreases, infiltration is enhanced, and the soil water availability is thus increased.

3.2.2 Types of drainage systems

Different systems are used in irrigated lands and rain fed areas; the type depends on the hydraulic characteristics of the soil, the slope, and the land use.

Furrows and shallow ditches: In flatlands, the overland flow can be discharged directly into a shallow ditch running parallel to the collector drain if crops cover the ground surface and there is sufficient slope. To protect the bank from erosion, water should be discharged from this ditch into the collector drain through several short pipes (Figure 3-27).

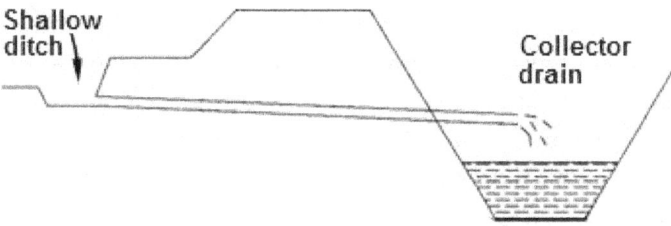

Figure 3-27: Shallow ditch connected to a collector drain

In lands irrigated by surface irrigation, better drainage conditions are obtained if furrows are used because, in addition to maintaining the plants in a relatively high position, they transport the excess water toward the outlet ditch,

Beds and dead furrows: This system is suitable to drain the excess rainfall from clay soils with low infiltration rate, especially if grassland is the major land use. The system consists of beds provided with transversal slope, which are constructed by ploughing, and dead furrows running with the ground longitudinal grade.

Excess water flows through the bed surface although some interflow also can occur; the total flow is collected in the dead furrows, which transport it toward crossable surface drains, which convey it toward the field outlet (Figure 3-28)

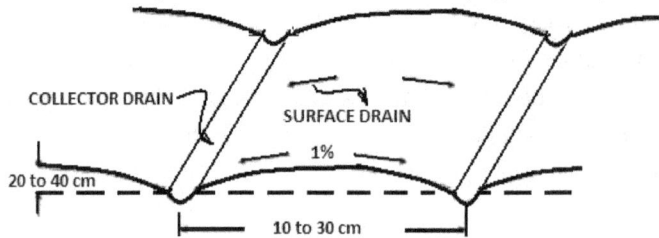

Figure 3-28: Layout of a bed drainage system for flatlands.

Parallel surface drains: For field crops, a system of parallel surface drains is generally more appropriate. Once the land has been graded and smoothed with the appropriate slope, surface runoff flows through furrows and is discharged into surface drains, which transport the drainage water toward a shallow ditch from which it is discharged into an open collector drain. Usually, the surface drains have lengths up to 250 m and are spaced from 100 to 200 m, according to the ground slope, land use, and land consolidation of the project area.

Parallel ditches: If, in addition to surface drainage, some amount of subsurface drainage is required, instead of surface drains, a system of parallel shallow ditches would be the most appropriate.

Surface drains in undulating and sloping lands: In undulating lands, surface runoff can be collected through a surface drain running along the low-lying valley, as shown in Figure 3-29. The drain depth depends on the amount of subsurface drainage required: If it is not relevant, shallow drains are appropriate, otherwise, deeper open ditches are preferred. If the valley is broad enough and there is the need to protect the bottomlands from the surface runoff flowing from the adjacent highlands, interceptor drains must be constructed in the valley at the foot of the slope.

Figure 3-29: Drainage system for undulating lands.

System consists of collector drains, which can be either pipes or open ditches; main drains, which generally are open canals; and auxiliary structures. Drainage disposal can be made effective through gravity outlets, tidal gates, or, in polder areas, through pumping stations. In areas receiving seepage from adjacent lands, interceptor drains are necessary to divert the outside flow. In polder areas, the agricultural area must be protected from flooding by means of embankments. In this way the project area forms a hydrologically independent unit.

The major functions of the main drainage system are to convey the drainage water toward the drainage outlet by the shortest route, to store excess rainfall during the rainy season to reduce the discharge at the outlet, and to maintain appropriate water levels during the dry season for controlled drainage management.

Reuse of the drainage water

An additional practice to reduce the volume of water to be disposed of is to reuse the drainage water to irrigate salt-tolerant crops and forest trees. The long-term feasibility of reuse increases with regional scale as opposed to on-farm-scale reuse because a system for the collection and redistribution of drainage water is required to reduce the need for storage reservoirs. Direct reuse by pumping from the open drains is a common practice in areas where water resources are scarce at some period during the crop season.

Disposal of drainage water

Drainage water usually is discharged directly to surface freshwater bodies such as rivers and less frequently into lakes, or directly into the sea, which under natural conditions is the permanent receptor of salts. The water flow of the river, the interest of downstream users, and the concentration of salts and pollutants in the river and drainage waters, which may vary during the year, determine the volume of drainage water to be discharged.

In coastal areas, where irrigation returns flow discharge into tidal rivers, salt contamination is not a severe problem but, if direct disposal into rivers is to be made, strict control of chemicals is necessary. In inland areas, several options exist to minimize the degradation of the quality of water bodies as a result of drainage water disposal: drainage canals to convey the water directly to the sea, evaporation ponds, constructed wetlands, and injection of drainage water into deep wells.

3.3   Flood control

Engineers and meteorologists are primarily concerned with the control of flood which occurs in head water areas and on major tributaries. The total flood losses increase with the size of the drainage area but the losses per unit area decreases. Drainage of flood is primarily on agricultural land. The principal headwater flood control measure includes:

1. Proper watershed management and storage of water in small reservoirs.
2. Proper watershed control measure reduce runoff and they also result in corresponding decrease in soil loss

Headwater flood control programme are concerned with such related activities as drainage, irrigation, gully and stream bank erosion control and land clearing.

Further reading

Amy Vickers (2001) "Handbook of Water Use and Conservation" WaterPlow Press.

De Jager J. M. and J. A. Kennedy. (1996). Weather-based irrigation scheduling for various farms (commercial and small-scale). Irrigation Scheduling: From Theory to Practice, eds. Smith, M., et al., FAO Water Reports, Vol. 8, pp. 33–38. Rome International Commission on Irrigation and Drainage and FAO.

Grattan, S. R., and J. D. Rhoades. (1990). Irrigation with saline groundwater and drainage water. Agricultural Salinity Assessment and Management, ed. Tanji, K. K., pp. 432–449, ASCE, New York.

Hill, R. W., and R. G. Allen. (1996). Simple irrigation calendars: a foundation for water management. Irrigation Scheduling: From Theory to Practice, eds. Smith, M., et al., FAO Water Reports, Vol. 8, pp. 69–74. Rome: International Commission on Irrigation and Drainage and FAO.

Irrigation Training and Research Center, (2000). "Ag-Irrigation Management"

Kay, M. (1983). Sprinkler Irrigation. Equipment and Practice. London: Batsford Academic and Educational.

Keller, J., and R. D. Bliesner. (1990). Sprinkle and Trickle Irrigation. New York: Van Nostrand Reinhold.

Martin, D. L., E. C. Stegman, and E. Fereres. (1990). Irrigation scheduling principles. Management of Farm Irrigation Systems, eds. Hoffman, G. J., et al., pp. 155–203. St. Joseph, MI: American Society of Agricultural Engineers.

Pair, C. H.,W. H. Hinz, K. R. Frost, R. E. Sneed, and T. J. Schiltz. (1983). Irrigation (5th ed.), Arlington, VA: Irrigation Association.

Threadgill, E. D., D. E. Eisenhauer, J. R. Young, and B. Bar-Yosef. (1990). Chemigation. Management of Farm Irrigation Systems, eds. Hoffman, G. J., et al., pp. 749–780. St. Joseph, MI: American Society of Agricultural Engineers.

Westcot, D. W. 1988. Reuse and disposal of higher salinity subsurface drainage water: A review. Agric. Water Manage. 14: 483.

Part 2

AGRICULTURAL MATERIALS
AND PROCESSING

CHAPTER 4

Agricultural Materials and Harvest

4. Introduction

The behaviour of most agricultural materials deviates essentially from that of the generally known elastic materials. The importance of economical production of agricultural material, especially crops and animal products serving as base materials for food stuffs, and of their technological processing (mechanical operations, storage, handling etc) is ever increasing. During technological processes agricultural materials may be exposed to various mechanical, thermo electrical, optical and acoustic (e.g. ultrasonic) effects. To ensure optimal design of such processes, the interactions between biological materials and the physical effects, acting on them, as well as general laws governing the same must be known.

It is practically impossible for machinery engineer to design machines for the handling and processing of agricultural materials without an understanding of the engineering properties of these materials. Some of these properties will be examine in subsequent sections below.

4.1 Engineering properties of agricultural materials

It has become entirely apparent that without a powerful and efficient agricultural technological-tools such as machines, devices, installations, automation; robotics etc -it will be impossible to secure food for the rapidly growing world population. This can only be made possible by obtaining the knowledge of material-machine interaction. In obtaining knowledge of the interaction of machine and material, it is necessary to explore the results and methods of agro-physics.

Agro physics is the application of physics to agricultural technological process. Agro-physics is a discipline which is undergoing development and becoming an important part of the theoretical foundations of agricultural equipment. Its content include the investigation of the physical properties of agricultural materials (plants, animals, products) and of the physical processes to which these materials are subjected under various conditions created by existing, future, or modeled technological procedures.

The following material properties are critical in studying the behavior of agricultural material under these technological processes.

Physical properties of agricultural materials

Geometric properties such as the shape, dimensions, volume, surface area, roughness of surface, porosity etc are some of the parameters which are important in many problems related to the design and function of a particular machine or to the analysis of the behaviour of the products under processing.

Shape and dimensions of product must be known before the analysis of the process of refrigeration of fruits, separation of seeds, grains and other unwanted materials when using technical, pneumatic and electro- static installations for separation and also for the optimal state of technology of granulated feed stuff or of the housing system for farm animals in a shed and in projection of transport of solid material by air or water.

Mechanical properties

Mechanical properties such as tensile strength, bending and shear strength in quasi-static, and impact stress and modulus of elasticity are important, and in certain instances quite indispensable technical data in the study of the functioning of grinding equipments, cutting, crushing, milling, squeezing, and conditioning.

The study of the resistance of products against damage in planting, sowing, harvesting and further manipulation and storage are important mechanical properties. It also applies to eggs, meat; bones and other parts of animal bodies. In rational designs and construction concerning movement of materials in harvesting, determination of grain and silage pressure against the walls of containers and silos, friction coefficients at rest and in motion are necessary.

Compressibility, elastic recovery, internal friction coefficient, elasticity of feeding stuffs and of silage is important characteristics in the study and determining methods and constructions of machines for compressing, bailing briquette. Similar properties of soil are considered in the construction of roughs and other tillage equipment. In animal husbandry, the analysis of effect of the teat-cup rubber upon the teat in milking machines needs the knowledge of rheology.

Energy to pick, and plant in planting machine, pull the potato tops and to separate spouts from tuber just as the force required to separate a tomato fruit from the plant requires the knowledge of the force and energy. The study of excited vibrations of a layer of cereal grain mass is necessary in considerations concerning vibratory threshing mechanisms.

Aerodynamic and hydrodynamic properties

In pneumatic and hydraulic transport and separation of material, these properties are useful. Resistance coefficient is used in calculation of critical speed of an object in liquid. Sedimentation process and rheology of liquid feedstuffs, liquid chemicals for plant protection, of dung-water, are very important. In design of pipe transport, pumps etc, these properties are required.

Volume characterizing airflow through a layer of grain or stems can be used in the construction of silos, and storehouses, in ventilation and drying. The behaviour of grains in the separating unit of a combine harvester, in a seed-clearing plant and in a pneumatic, conveyor system uses the aerodynamic properties of seeds. This property has been used extensively in dispersal of granular materials from agricultural aircrafts.

Thermal properties

A number of agricultural products of plants or animal origin are subjected to all sorts of thermal properties, which include heating, cooling, and freezing. These processes are not only applied in conservation or preparation of feedstuffs but also in activating hard non-germinating seeds under controlled heating. In fruits and vegetables, the activity of enzymes and microorganism causing determination is commonly limited by low temperature. Heating and cooling farm crops and products is achieved by heat transmission such as thermal characteristics as specific heat, thermal conductivity and heat capacity, heat transfer, radiation intensity together with density, shape and dimensions are among the basic data in the construction of the corresponding installations and in process designing.

Acoustic properties

Acoustic properties can be utilized in the study of toughness of whole fruits and thus to suggest the correct time of picking from the tree without damage and to know the connection of resonant frequencies at different modes of vibration with the size and sort of the fruit-Acoustic vibration can be used in artificial pollination of tomato in glass house. Acoustic waves can be used against insect activity. Egg fertility is tested by ultrasound. Ultrasound can detect thickness of layer of fat under pig's skin, which can be used for correct feeding and in the breeding of new lines with more flesh than fat.

Electric and magnetic properties

Electric properties of agricultural materials, which are important in processing and manipulation, are electric conductivity, capacitance dielectric properties and behaviour in magnetic field in general. Grain moisture content is determined on the basis of electric conductivity. Electric resistance measuring method was applied in the measurement of length pattern of cotton fibers to measure fiber fineness. Principles of electric separation are utilized in cleaning of grain crops. Measurement of the conductivity of milk can supply information during the process of milking, quality of milk and the state of health of cow.

Dielectric heating using microwave technique ensures unit heating of material and rapid temperature rise. The knowledge of electric properties of stems of fodder crops was the basis for the construction of a discharge disintegration of plant tissues both as equipment facilitating mechanical dehydration of green fodder. In mechanics of animal husbandry, the construction of electric fences must be with a thorough knowledge of electric properties of processes taking place with the contact of animal tissue with a charged conductor with regard to electric properties of the earth potentials.

Optical properties and the application of ionizing radiation

 Light transmittance and reflectivity at different wave lengths of agricultural products has been used in recent years in electronic separation, sorting and determination of ripeness, surface colour and generally in the study of external and internal characteristics of fruits, vegetables and pulse crops. An automatic tomato and lemon-sorting machine is used which sorts fruits into groups. Similarly, photoelectric equipment with optical filters is used in weeding and singling machines for sugar beet cultivation. Photoelectric sensor had been used for distinguishing between plant material and soil. This detects the presence of a seedling in a tungsten halogen light beam scanning over the soil sample. Photoelectric seed counting detector detects accurately a broad range of seeds such as tobacco and corn, approximately ±1.54%.

4.2 Rheology

Rheology is the study of mechanical deformation and flow resulting from the application of forces, with time effects taken into account. It is concerned with the relationships between time dependent stresses and deformations, creep and stress-relation and the study of viscosity. The characteristic structures of biological materials necessitate the introduction of certain concepts and definitions, which are not usual in the mechanics of common elastic bodies. These are:

Bearing load: This is a compressive load curve obtained when the applied stress decreased or remain constant with increase in material deformation. This is an indication of initial cell rapture and it is also a measure of cell sensitivity to damage.

Rupture point: A point on the stress strain or force-deformation curve at which the axially loaded specimen ruptured under load. Rupture indicates failure in the macro structures.

Rigidity: Rigidity is characterized by the tangent to the initial linear section of the stress-deformation curve. This is the modulus of elasticity.

Degree of elasticity: the ratio of the elastic to the total deformation when the material is loaded. The total deformation is the sum of the plastic and elastic deformation.

Plasticity: Is the capacity of material for taking plastic or permanent deformation.

Toughness: The work required to cause rupture in a material (mNm^{-3}). This is identical to the area under the stress-deformation curve.

Resilience: The capacity of a material to store strain energy in the elastic range. This is deformation work.

Mechanical hysteresis: This is the energy absorbed by a material in a cycle of loading - unloading.

Energy recovery: This is the ratio of the energy recovered on unloading to the energy invested in loading.

4.3 Crop harvest systems

Harvesting simply mean the removal of an entire economic product, or its economic parts, after maturity from the field. The economic product may be grain, seed, leaf, root or the entire plant. Portions of the stalk that are left in the field are called stubble. The time of harvesting is determined by the degree of maturity. With cereals and pulses, a distinction should be made between maturity of stalks (straw), ears or seedpods and seeds, for all that affects successive operations, particularly storage and preservation.

Crop maturity

Crops are harvested either at physiological maturity or at harvest maturity. Physiological maturity is the stage at which translocation of food matter to the economic part stops. That is, after this stage, no further increase of dry matter occurs in the economic part. Moisture content in cereal grains is very high during the milking stage (when the grain is barely formed) and it gradually decreases (from 40 to 20 percent).

At this stage, translocation of carbohydrates is stopped due to the formation of a hard ring (known technically as an abscission layer) at the neck of the grain (rachis). Harvest maturity occurs approximately 7 days after physiological maturity (depending on the prevailing weather conditions). The important process during this period is loss of moisture from the plants. The general symptoms of harvest maturity are yellowing of leaves and drying of grain pods. Physiological maturity harvesting is advisable only if the field needs to be vacated for the next crop. Otherwise, harvest maturity is recommended.

Time of harvest

Timely harvesting is essential. Readiness for harvesting is calculated on the basis of the moisture content of grain. Early harvest gives immature grains with high moisture content. Such grain is difficult to store because it is susceptible to pests. In such cases, the yield, the quality of grain, as well as the germination potentials are reduced. Late harvesting, on the other hand, may cause the grain to shatter. In addition, crops harvested at very low moisture may lead to damage of seeds while threshing.

The mechanization of harvesting and subsequent manipulation (packaging and handling) operations could have unfavourable consequence (damage) on the products when not adequately handled. The quality of the product is adversely affected followed by rapid spoiling and deterioration. Agricultural materials are damaged most frequently by impact during harvesting. Experience indicated when agricultural material is ready for harvest subsequent processing.

Handling of agricultural products

Handling is the final stage in the process of producing high quality fresh product. Good product quality is attributable to appropriate production practices, careful harvesting, and proper packaging, storage, and transport systems. Being able to maintain a level of freshness from the field to the dinner table presents many challenges. Production practices have a tremendous effect on the quality of fruits and vegetables at harvest and on postharvest quality and shelf life.

The inherent quality of produce cannot be improved after harvest, only maintained for the expected window of time (shelf life). Part of what makes for successful postharvest handling is an accurate knowledge of what this window of opportunity is under your specific conditions of production, season, method of handling, and distance to market.

Crops destined for storage should be free as much as possible from skin break, bruises, spots, rots, decay, and other deterioration. Bruises and other mechanical damage do not only affect appearance, but provide entrance to decay organisms as well.

In organic production, farmers harvest and market their produce at or near peak ripeness more commonly than in many conventional systems. However, organic production often includes more specialty varieties whose shelf lives and shipping traits are reduced or even inherently poor. As a general approach, the following practices can help maintain quality:

1. Harvest during the coolest time of day to maintain low product respiration.
2. Avoid unnecessary wounding, bruising, crushing, or damage from humans, equipment, or harvest containers.
3. Shade the harvested product in the field to keep it cool. By covering harvest bins or totes with a reflective pad, you greatly reduce heat gain from the sun, water loss, and premature senescence.
4. If possible, move the harvested product into a cold storage facility or postharvest cooling treatment as soon as possible. For some commodities, such as berries, tender greens, and leafy herbs, one hour in the sun is too long.
5. Do not compromise high quality product by mingling it with damaged, decayed, or decay-prone product in a bulk or packed unit.
6. Only use cleaned and, as necessary, sanitized packing or transport containers.

These operating principles are important in all operations but carry special importance for many organic producers who have less access to postharvest cooling facilities.

4.4   Methods of harvesting agricultural materials

The selection of a harvesting procedure for particular agricultural crop depends on the characteristics of the product. Harvesting may be done either manually by hand picking or mechanically using aids and equipments.

4.4.1   Manual harvest by hand picking

A large amount of harvest is still done by hand; this labour may represent about half of the cost of growing a particular crop. The sickle is the most important hand tool in manual harvesting of grains such as rice. A knife may also be used for plants with woody stems.

In some countries fruit is harvested by hand, placed onto straw on the ground under the trees, hand-sorted, and packed into containers before leaving the orchard. Under these conditions consistent quality control is difficult to achieve among orchards, but fruit may sustain less handling damage. Fruit are easily damaged at harvest time, so care is required.

Figure 4-1: Picking mango fruit

Hand picking tools

Majority of crops are harvested by hands using secateurs, clippers, knives or diggers. Some fruits such as citrus, grapes and mangoes, need to be clipped or cut from the plant (Figure 4-2).

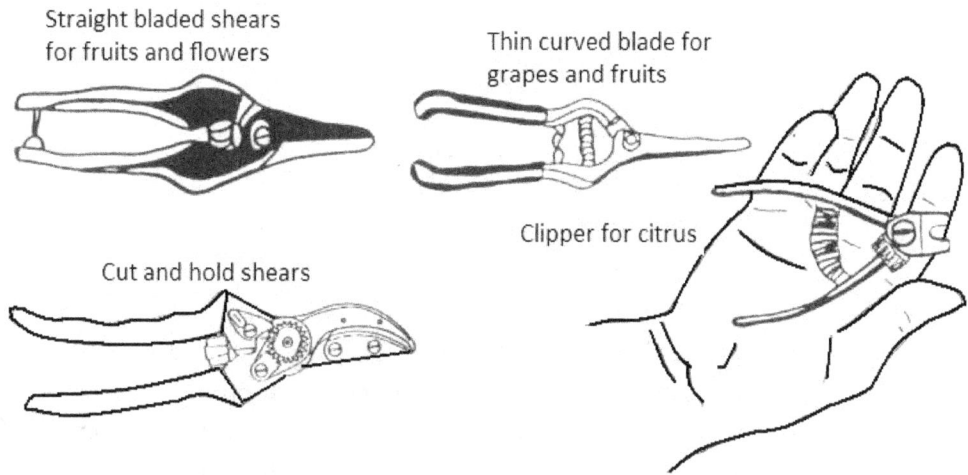

Figure 4-2: Hand picking tools

Clippers or knives should be kept well sharpened and clean. Peduncles, woody stems or spurs should be trimmed as close as possible to prevent fruit from damaging neighboring fruits during transport. Care should be taken to harvest pears so that the spurs are not damaged. Pruning shears can be used for harvesting fruits and some vegetables.

Figure 4-3: Harvesting with secateurs

Picking bag: Fresh fruits harvested by hand were dropped into suitable picking bags hung on shoulders to avoid damage. Cloth bag with openings on both ends can be easily worn over the shoulders with an adjustable harness (Figure 4-4). In case metallic buckets are to be used for harvesting, fitting cloth over the opened bottom can reduce damage to crop. Fitting canvas bags with adjustable harnesses or by simply adding some carrying straps to baskets also helps to reduce handling losses.

Figure 4-4: Picking bags

Picking poles and catching sacks: These tools can be easily made by hand. A long pole attached to a collection bag, allow the harvester to cut catch produce growing on a tree without climbing on tree. The collection bags can be hand woven from strong cord or sewn from canvas. The hoop used as the collection bag rim and sharp cutting edges can be made from sheet metal, steel tubing or recycled scrap metal.

Figure 4-5: Picking pole

Tripod ladders: A ladder with three legs is very convenient and more stable than a common ladder (Figure 4-6). A ladder helps harvesting crops such as mango, pears, peaches, plums without damaging tree branches.

Figure 4-6: Tripod ladder

Advantages and disadvantages of manual harvesting

The primary advantages manual harvesting includes:

a. Harvesting of fruit or vegetable can be done at appropriate maturity.
b. The produce will suffer minimum damage.

Disadvantages includes;

a. It is a time consuming process.
b. More labour is required during harvesting season.

4.4.2 Mechanical harvest

Mechanical harvesting is done with the help of harvesting aids and combines. Combines perform several operations such as cutting the crop, separating the grain from the straw, cleaning it of chaff and transporting it to the storage tank.

Various mechanical means have been developed for harvesting of fruits and vegetables. Common among them are tree shaking, mechanical aids and mechanical harvesters.

Tree shaking: Harvesting involves shaking the tree or cane by mechanical vibration and catching the detached fruit underneath in a large blanket or net. However, these systems can cause significant damage to the crop and are generally only suitable for fruit to be used for processing. There are difficulties if the fruit on the tree do not all ripen at the same time.

Mechanical aids: Mechanical aids are improvements on the tree shaking and hand picking methods of harvesting fruits. Mechanical aids are available for harvesting in the form of gantries, picking ladders, and in some cases mobile conveyor systems, but more commonly the picker collects the fruit into a small holder such as a picking apron or bucket, which holds not more than 15 kg of fruit. Mangoes, papaya, apples, and fruit grown in tall trees can be harvested using picking poles. The fruit is separated from the tree by a sharp cutting edge on the end of the pole and falls into a net just under the cutter.

Mechanical harvesters: Mechanical harvesters have been developed for many crops including apples, strawberries, blackcurrants, blueberries, cherries, and raspberries. Some fresh vegetables are harvested by mechanical means. These include peas, beans, tomatoes, sprouts, and root crops, particularly if the product is for processing.

Advantages and disadvantages of mechanical harvesting

The primary advantages manual harvesting includes:

a. The produce can be harvested at a faster rate.

b.   Less manpower is required as compared to hand harvesting.

Disadvantages

a.   Damage can occur to crops.
b.   Not suitable for marketing of fresh commodities

Mechanical damage

Mechanization of various harvesting and subsequent manipulation of operation has an unfavorable consequence in that it leads to an increase in damage to the material processed. The quality of the product will be lowered and in many cases followed by rapid spoiling and complete deterioration.

Reduction of mechanical damage in agricultural materials can be achieved by:

1.   Reducing the cutting forces to the lowest level during designing
2.   Breeding of improved quality material which can withstand great loads
3.   Harvesting when materials are matured enough to resist mechanical strength of equipment.

Causes and effects of damage

Damage may appear in very diverse forms, from hair cracks to total rupture. Damage could cause such effects as decrease in germination capacity of cereal grains and increases the rate of oxidation during storage. Mechanical picking of fruit implies significant mechanical damage. On shaking, the impact against the tree branches, other fruits and finally the catching surface causes tissue beneath to deform leading to browning and spoilage. If the deformation passes the biological yield point, tissue will brown within a short time and be spoilt. In certain cases, browning under the skin is not visible from outside (e.g. in pears)

Certain fruits e.g. tomato and cherries fall off without their stems attached on shaking the tree. Juice appears on the scar left on fruit, representing a loss on one hand and promoting deterioration on the other. Significant damage is recorded during loading and unloading of root bulbs. Damage results from impact, bruising and cuts.

It is therefore obvious that damage is generally caused by static and dynamic external forces. Mechanical damage resulting from internal forces is cursed by physical variations taking place inside a product, for example variations of the temperature and moisture content, or chemical and biological variations. Cherries and tomatoes have skin cracking due to an increase in internal pressure.

Establishing and measuring damage

The methods of assessment established in practice are described below, from which the most appropriate to given circumstance must be selected. The main forms of appearance of mechanical damage are as follows:

Abrasion: The skin of a product is damaged or partly separated from the tissues beneath. Abrasion is sometimes hardly visible after harvesting but will become apparent after few days of storage.

Bruising: Damage to plant tissue occurs as a result of external forces causing physical changes.

Cracking: This is limited to cracking of skin or tissue due to impact or pressure, without causing the product to fall apart into several pieces.

Cutting: This is the penetration of a sharp tool into the product without any significant crushing effects.

Puncture: Caused by pointed needle-type tools, plants stems, or thorns penetrating the surface of a product and the tissue beneath.

Shatter cracking: Multiple cracks starting radially from the point of impact.

Skin cracking: Cracks restricted to the outer skin alone.

Splitting: A product divide into several parts

Tearing: Usually caused by stem ends i.e. when the skin of a fruit is thorn on removing the stem.

Swell cracking: Caused by an increase in internal osmotic pressure.

Distortion: Distortion concerns changes of form caused by loads acting on a product.

Sensitivity to damage

Sensitivity to damage is influenced by numerous factors which include

1. Physical state of the material e.g. temperature, moisture content etc.
2. Biological state of the material e.g. state of growth, ripeness etc.
3. Load characteristics such as static, dynamic, oscillating, loading rate etc.

Effects of various parameters on sensitivity to damage

Sensitivity to damage is influenced by numerous factors. One group of factor/parameter is concerned with the physical and biological state of the material e.g. temperature, moisture content, stage of growth, and ripeness, while other is related to the load characteristics e.g. static, dynamic, oscillating, loading etc.

In most cases, temperature greatly effects the mechanical properties of agricultural products and thereby their sensitivity to damage. With variation of temperature, the tugor pressure of cellular material and together with it the elasticity both vary.

Effects of mechanical harvest on crops

Damage during mechanical harvesting can lead to a number of undesirable changes in produce. Physical damage caused by mechanical harvesting methods may lead to:

1. Water loss
2. Increased respiration rate
3. Initiation of ethylene synthesis
4. Production of undesirable colours (browning)
5. Penetration of plant tissues by micro-organisms (both plant and food borne human pathogens)

Mechanical harvest machines and operations

Mechanical harvesting is currently one of the most active fields, of research for agricultural engineers and crop scientists. It requires the geneticists to breed crops of nearly equal size that mature uniformly and that is resistant to mechanical damage. Several machines have been designed and manufactured for the harvesting of agricultural products in the past and now. Special combines have been designed for harvesting particular crops ranging from grains to root crops and tree crops.

Figure 4-7: Combine harvesting

The combine reaps two to nine rows at a time and is usually equipped with an 8-10 HP engine. This is a special combine designed for the harvesting of grain crops such as rice, corn, millet, etc. This machine cuts the crops, feed the crop into the cylinder, threshes the seed from the seed head, and separates the seed or grain until it is dumped into the truck or trailer.

Figure 4-8: Harvesting with tractor

The conventional combine

Figure 4-9 illustrates the operating principles of a typical conventional grain combine. Crop is fed tangentially into a cross-mounted cylinder-concave assembly. Threshing occurs largely by impact of the cylinder bars on the incoming crop, while considerable separation occurs through the open grate concave. Separation of the remaining grain from the straw is accomplished with straw walkers, while a cleaning shoe, with chaffer and sieve, is used for scalping and final cleaning.

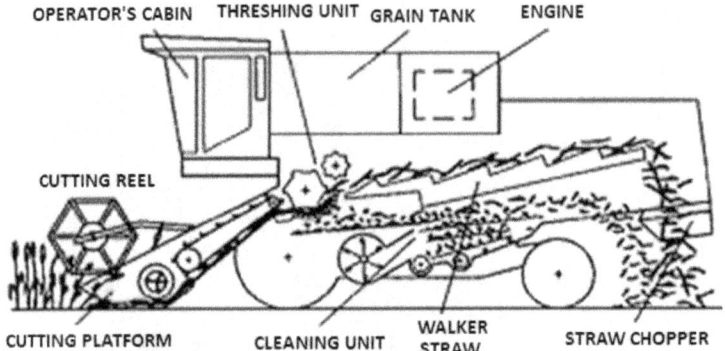

Figure 4-9: Conventional combines functional units

Operation of combine harvester

The pickup reels raises the stalk and position them to be cut by the cutter bars, transport them to the feeder conveyor by the platform augers, the grain heads move into the cylinder concave where threshing is done. Separation of grain is achieved in the straw walkers through shaking action as they move out of the combine. The grains are cleaned in the chaffer while clean grain falls through the openings between the louvers of the chaffer. The chaffer and sieve are referred to as cleaning shoe.

Components of combine harvesters

The main processes in a modern combine harvester involved gathering, cutting, conveying, threshing, separating, cleaning, and materials handling. Figure 3-9 shows the main elements of a conventional combine.

The header (cutting platform) divides, gathers and cuts the crop with a reel and cutter bar then a screw auger conveys the crop into the feeder house for presentation to the threshing unit.

Figure 4-10: The combine header

The platform auger: This gathers the crop head to the center of the platform where the feeder conveyor delivered it to the cylinder for threshing (Figure 3-11). The feeder conveyor receives the crops from the platform and conveys it for threshing.

Figure 4-11: Platform auger

The threshing unit of a combine consists of threshing drum and cylinder concave. Threshing is accomplished by a combination of impact and rubbing action. The tasks of the threshing units include the separation of the grain through the concave and transferring the straw to the straw walker or separating section.

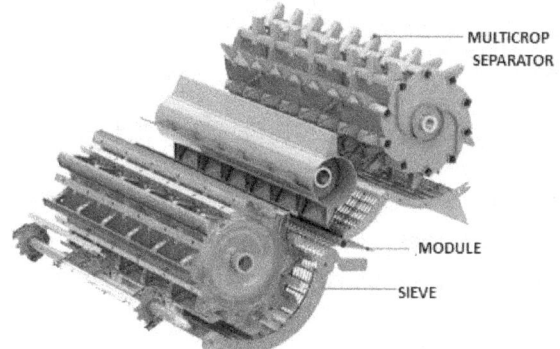

Figure 4-12: Threshing unit

The separator unit or straw walker separates the remaining grain from the straw. Straw walkers take up a lot of space but have a low power requirement for grain-straw separation. Four to eight long, permeable channel sections with a width of 0.25–0.35 m and a length up to 6 m are mounted on two crankshafts that rotate at approximately 200 rpm.

Figure 4-13: The separator unit or straw walker

Straw bounces on top of the channel sections; grain and some chaff are sifted down and separated from the straw. Four to five steps assist this operation by further loosening the straw layer. The separated grain is conveyed to the grain pan of the cleaning unit by a set of augers or by a return pan below the channel sections.

The cleaning unit does the final separation of grain from the chaff and broken straw pieces delivered by the concave and straw walker. The clean grain is conveyed into the grain tank. Unthreshed grain heads or pieces of cob (tailings) are returned for re-threshing.

Figure 4-14: The grain cleaning unit

Frame structure: All the components mentioned above were mounted on to a solidly built frame network which gives it exceptional strength and rigidity and enabled it to withstand the toughest and most demanding field conditions.

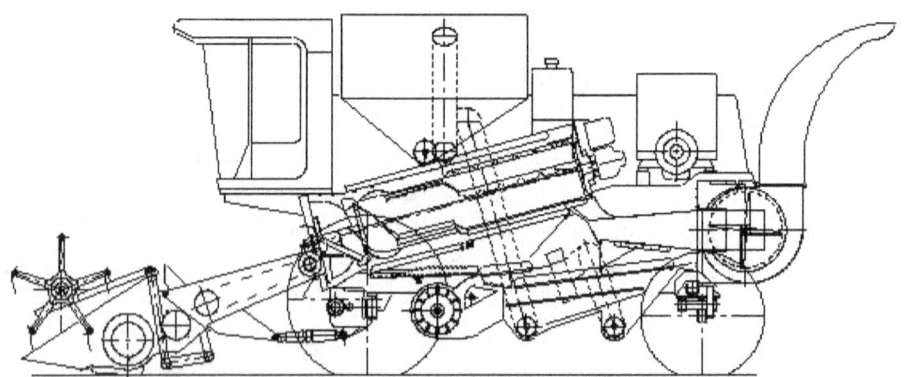

Figure 4-15: Combine frame and components

Types of combine harvesters

1. *Self propelled combines*: This provide power for moving the machine through the field and during operation. A constant source of power is delivered to the threshing, cleaning and separating units.

2. *Hill side combines*: This is designed for harvesting crops in rolling terrains with an automatic leveling system.

3. *Pull type combine:* This type has similar feature with self-propelled combines only that they are tractor drawn.

4. *Special combines*: These are designed for harvesting a particular crop e.g. okra harvester, rice or soybean combines etc.

Figure 4-16: A self-propelled combine

New types of combines

Along the line, the improvement of working technologies in agriculture was searched in order to increase the working capacity and the output of harvest machines. Important progresses have been made lately in cereal harvesting by using the integral technology due the ever increasing requirements imposed on this process (such as minimum losses, reduced power consumption, optimum harvesting time).

Figure 4-17: A self-propelled corn harvester

The first models of self-propelled cereal harvesting machines built were those having a tangential thresher, but after 1970, as a result of the continuous demand of the farmers for increasing the output, harvesters with axial thresh mechanisms were developed permitting both the raising of the working capacity and the rate of the seeds separated during the threshing process.

A number of new types of combines, incorporating different threshing and separating concepts have recently been introduced and these include.

a. *Western Roto thresh* used a different principle of grain-straw separation than either conventional combines or the newer axial type combines. The Western Roto-Thresh used a conventional cylinder-concave assembly for threshing and initial separation. Final separation of grain from the straw was accomplished with a large diameter, slow speed, perforated separating drum. A conventional cleaning shoe, combined with an aspirator system was used for scalping and final cleaning.
b. *Sperry New Holland:* Sperry New Holland uses two longitudinally mounted, axial threshing and separating rotors. Threshing occurs in the threshing concaves at the front of the rotors, while separation of grain from straw is accomplished along the full rotor length. A rear beater-grate assembly performs final separation. A conventional cleaning shoe is used for scalping and final cleaning. Several different sizes of this combine are currently being produced.
c. *International Harvester:* International Harvester uses a single, longitudinally-mounted axial flow threshing and separating rotor. Threshing occurs at the front section of the rotor, while separation of the grain from straw is accomplished along the full rotor length in both the threshing and separation concaves. A rear beater aids in straw discharge. A conventional cleaning shoe is used for scalping and final cleaning. Several different sizes of this combine, including a pull-type version, are being produced.
d. *White:* White also uses a single, longitudinally-mounted axial threshing and separating rotor. Threshing occurs at the front section of the rotor, while separation of grain from straw is accomplished along the full rotor length in both the threshing and separation concaves. A conventional cleaning shoe is used for scalping and final cleaning.
e. *Allis Chalmers:* Allis Chalmers use a different design than the previous three combines. The threshing and separating rotor (cylinder) is mounted crossways, with crop fed tangentially into one end of the rotor. Threshing and separation occurs along the full length of the rotor as the crop spirals sideways along the rotor. A paddle and impeller assembly discharges the crop from the outlet end of the rotor. A conventional cleaning shoe, combined with accelerator rolls and a high velocity air blast, is used for scalping and final cleaning. Several sizes of this combine are available.

Fruit combine

Virtually all previous work on the mechanical harvest has been concentrated on one form or other of   the shake - and – catch principle. Although high rates of fruit removal can be achieved by tree shaking, the problem of fruit damage during and subsequent to removal has not yet been overcome. Combining action system was employed in the harvesting of apple to prevent damage. The effective harvest of okra could be realized if the cutting unit were properly locate relative to the okra plant stalks, and if adequate control were exercised in order to avoid damage to immature fruits of the plant.

Figure 4-18: Mechanical fruit harvester

Factors affecting machine performance

Several problems often affect the optimal functioning of machines such include:

1. Very short period of straw cereal harvesting (8 to 10 days optimum),
2. Unfavourable climate conditions during the harvesting period,
3. Fallen plants heads embedded in weeds or non-uniformly grown plants leading to important grain losses etc.

The harvesting machines being so designed that they should perform the technological process to a high quality level. In order to do that, the harvesting machines should comply with a series of requirements such as:

1. Straw losses not to exceed 5%;
2. Percentage of broken grains should be less than 2%;
3. Purity of the collected grains should be over 98%;
4. Width of the cut and threshed straw and ear swath should not exceed 1, 200 mm etc.

Technology research and innovations have provided solutions which are being adopted by various companies for raising the harvester output and increasing the percentage of the seeds separated by the thresher.

4.5 Harvest effect on postharvest quality

Great care must be taken during harvesting of perishable fruits and vegetables to avoid physical damage. Any mechanical damage that occurs at harvest, during movement of product to the pack house, or through grading and packing lines will result in enhanced respiration, increased ethylene production, water loss and increased susceptibility to infection by postharvest pathogens, all of which can induce rapid deterioration and loss of quality.

A number of simple but effective steps can be taken to reduce physical damage from occurring during this phase of the harvesting and handling system. These include careful handling of the product at all stages of the operation, good sanitation and hygiene with all equipment, maintenance of packing equipment to prevent excessive drops onto hard surfaces, and padding of all machinery surfaces on which products may impact.

The deterioration of fruit after it has left the tree depends on one or some of the following factors:

1. Growth and activities of micro organisms.
2. Activities of natural food enzymes
3. Insects, pesticides and rodents
4. Temperature; both heat and cold
5. Moisture and dryness
6. Air and in particular oxygen
7. Light
8. Time of harvest.

Fruit damage at harvest

Fruits are easily damaged at harvest time, so care is required. Mechanical aids are available for harvesting in the form of gantries, picking ladders, and in some cases mobile conveyor systems, but more commonly the picker collects the fruit into a small holder such as a picking apron or bucket.

Mangoes, papaya, apples, and fruit grown in tall trees can be harvested using picking poles: Fruit is separated from the tree by a sharp cutting edge on the end of the pole and falls into a net just under the cutter. In some countries fruit is harvested by hand, placed onto straw on the ground under the trees, hand-sorted, and packed into containers before leaving the orchard.

Mechanical harvesters have been developed for many crops including apples, strawberries, blackcurrants, blueberries, cherries, and raspberries. Harvesting involves shaking the tree or cane by mechanical vibration and catching the detached fruit underneath in a large blanket or net.

4.6 Farm transport and machinery

Farm transportation is as old as human existence because even the early man who was only a gatherer still had to convey himself to the centers of food collection. Early form of transportation was mainly on-farm as the major activities were collection of water, crop gathering, animal hunting and related activities most of which were done within the neighborhood of the farmer and hence the distances covered were usually very short (Mijinyawa and Adetunji, 2005).

Farm transportation

A farm is a system or environment where agriculture and food production takes place. Transportation involves the movement of man and materials from one point to the other. Farm transportation is thus movement of agricultural food products and other related materials such as machinery, structure etc from the farm to the markets or from the source to the farm.

Farm transportation plays a key role in the agricultural and economic development of any nations as it provides access for extension agents to transfer new and improved agricultural technologies to the rural and farming communities, timely delivery of inputs to the farm and evacuation of harvests to the urban areas where they are mostly demanded. These ensure improvement in agricultural production, food availability in urban areas and improvements in the economy of the rural communities (Klatzel, 2000).

In Nigeria, farm transportation is particularly important, as most of the foods producing communities are located in the remote areas. It is for this reason that the existing farm transportation system should be improved upon and expanded within the economic and technical limitations of the end-users.

Means of farm transport

For reasons of economic and technical limitations, and the peasant nature of agriculture, farm transportation in many developing countries is substantially by land (road and rail) and water in the riverside and coastal regions.

There are a number of means through which we have transportation.

- *Road transport*: This is the commonest means of transport for agricultural materials
- *Sea transportation*: Products can be transported through sea or body of water
- *Rail transportation*: This is a means of transportation with attached coaches to carry heavy loads along iron rails.
- *Air transportation*: Moving materials in air is possible through the use of flying machines such as parachute, helicopters and aircrafts.
- *Animal transportation*: mostly common in Northern Nigeria to convey material or agricultural products e.g. Carmel horse donkey etc.
- *Ropeway transportation*: A means of transportation accepted as a means of communication e.g. metro line.
- *Pipe transportation*: This is mostly used for .dry and granulated materials and liquid viscous material.

Categorization of farm transport

There are two categories of farm transports viz:

On-farm transport: These are equipments for moving goods between field, store and household. On small farms this will include collection of wood and water for domestic purposes.

Off-farm transport: For the movement of goods between farm market. Loads are generally greater, distance longer.

Farm vehicles

There exist a wide range of low cost vehicles for moving, farm goods, which can be categorized as follows.

1. *Single equipment*: These are single unit carrying aids for agricultural materials such as:

- *Carrying aids for head*, shoulder, or back loading e.g. baskets, bags, sacks etc.

Head carrying aids   Bicycle

Figure 4-19: Some single equipment

- *Wheel barrows and hand carts*

Motorcycle   Wheel barrow

Figure 4-20: Some single equipment

- *Pedal driven vehicles* such as bicycles. Compared to other forms of transportation, the conventional bicycle is among the most efficient means of human locomotion.
- *Back of animals* such as ox, donkey, horse, cow etc as well as animal drawn carts.

2. *Intermediate equipment*: These are low power assisted transport equipments driven by human or animal assistance that attempts to merge the health and environmental benefits of a bicycle with the convenience of a motorized vehicle such as:
    - *Motorcycles and converted motorcycles*: These developed more power than the human locomotive bicycles. For instance, to travel one kilometer by bike requires approximately 5-15 watt-hours (w-h) of energy, while the same distance requires 15-20 w-h by foot, 30-40 w-h by train, and over 400 w-h in a singly occupied car (Justin, 2004).
    - Trailers for bicycles and motorcycles
    - Tricycles

3 *Advanced equipment*
    - Basic motorized vehicles
    - Dual-purpose agricultural transport equipment.

All these vehicles have different advantages and disadvantages in terms of load bearing capacity, suitability for route conditions, running costs, speed range and capital cost which enable them to meet a broad spectrum of transport requirements. Many low cost forms of transport are used only in certain local areas and remain unknown even in other areas.

Trailers

The large trailers pulled by semi tractors have their own rear suspension and wheels, the front of the trailer being supported by the fifth wheel on the tractor. Semi trailers also have folding supports under the front that are lowered when the trailer is detached from the tractor and parked. The brakes on the trailer's axles have air hoses that attach to the tractor's brake system, so the tractor and trailer brakes work together. Trailers have their own signal, tail, and brake lights, all of which are powered by the tractor's electrical system. Trailers come in many different designs, depending on the intended cargo. Enclosed or standard box-type trailers are used to haul a wide variety of goods and merchandise.

Figure 4-21: Double axel trailer

Double trailers are often used on roads that have sharp turns. Double trailers resemble two smaller trailers linked together and can maneuver through tight turns more easily than standard trailers can. Size and weight restrictions apply and vary from state to state. In the United States, tractor and single trailer combinations generally must be less than 16 m (53 ft) in length and are limited to a maximum weight of 36,000 kg (80,000 lb).

Figure 4-22: Agricultural trailers

Separate weight limits apply to trailers with single or tandem axles and to double trailers (Figure 4-22). Maximum trailer height and width are dictated by state law and vary from state to state. For most states, the maximum height is 4.11 m (13.5 ft), and the maximum width is 2.6 m (8.5 ft). In some states, trailers may be equipped with additional wheels and axles to carry heavier loads.

A special type of enclosed trailer is an insulated and refrigerated "refer" unit, used for transporting perishable food items. Refrigerated trailers have a small engine mounted on the trailer for powering the refrigeration system. This allows the refrigeration unit to run continuously, even when the trailer is parked or disconnected from the tractor.

Piggyback trailers are enclosed trailers designed to be mounted on railroad flatcars for cross-country transport. Some have their own wheels and suspension, while others are sealed containers that are lifted off and placed on a trailer chassis. Sealed containers are also used on special ships, called container ships, to transport goods overseas.

Figure 4--23: Piggyback trailer

Flatbed trailers are used to transport large objects such as construction equipment, industrial machinery, and oversized objects (Figure 4-24). Such trucks may be equipped with an Oversize Load warning sign and flashing lights, and may be accompanied by an escort vehicle to warn other motorists.

Figure 4-24: Flatbed trailer

Platform trailers are essentially large containers with open tops for transporting produce and grain (Figure 4-25). Special trailers are also designed for hauling livestock, automobiles, and beverages.

Figure 4-25: Flatbed trailer loaded

Tank trailers, known as tankers, are used to haul chemicals, milk, gasoline, and other liquids (Figure 4-26). Tankers, as well as other trucks that carry flammable or toxic products, must display special warning emblems to warn police and firefighters in case of an accident

Figure 4-26: Tanker trailer

Figure 4-27: Slurry tanker for liquid manure

Figure 4-28: Water distribution tanker

Trucks

Trucks play an important role in many farm operations. They are a prime form of transportation. They are also used to transport a large variety of materials and livestock. Trucks may be used outside regular working hours when required in critical farm operations such as harvesting a crop or tending livestock.

Figure 4-29: A truck bucket (trailer)

Figure 4-30: Loading a truck trailer

To minimize disturbances, trucks should be kept in good mechanical repair and not left idling when not in use.

Aircraft

Fixed wing aircraft are used to apply seed, fertilizer and pesticides to some commodities. Ensure that application by aircraft are specifically listed on pesticide labels and check with the local office of the Ministry of Water, Land and Air Protection for any further restrictions before application.

Figure 4-31: Aerial application of chemical

Aircraft attract the attention of neighbours and may be more of a disturbance than land based application methods. Helicopters may be used to dry cherries to prevent splitting and have also been used in frost protection. Ultra-light aircraft may be used to check on livestock on the range.

Further reading

Atiku A., Aviara N., and Haque M.. (2004). Performance Evaluation of a Bambara Ground Nut Sheller. Agricultural Engineering International: the CIGR Journal of Scientific Research and Development. Manuscript PM 04 002. Vol. VI. July, 2004

Bello R. S., 2012. Agricultural Machinery & Mechanization. 7290 B. Investment Drive Charl 7290 B. Investment Drive Charl createspace ISBN-13: 978-1456328764. https://www.createspace.com/3497673 (344 pages)

Benson E.R., Stombaugh T.S., Noguchi N. WillS., J.D and Reid J.F. (1998). An Evaluation of a Geomagnetic Direction Sensor for Vehicle Guidance in Precision Agriculture Applications. An ASAE International Meeting Presentation UILU 98-7011

Equipment Handbook (2003). Grain Inspection, Packers and Stockyards Administration U.S. Department of Agriculture Chapter 6 1400 Independence Ave., SW 10-04-96 Washington, D.C. 20250-3600

Hughie D. Kydd, and Humboldt Station (2003). Combine types. Prairie Agricultural Machinery Institute Lethbridge, Alberta Humboldt, Saskatchewan Portage La Prairie, Manitoba

John B. Liljedahl, Paul K. Turnquist, David W. smith, Makoto Hoki (1989). Tractors and their power units, 4th Ed. Published by von Nostrand Reinhold. NY.

Kanayama, Y., B. I. Hartman.(1989). Smooth local path planning for autonomous vehicles. In proc. Ieee international conference on robotics and automation, 3:1265-1270.

Lawrence, A. 1993. Modern inertial technology, navigation, guidance, and control. Springer-Verlag New York, Inc., New York, NY.

Mark Hanna, 2002. Estimating the Field Capacity of Farm Machines. File A3-24 Iowa State University Extension Service hmhanna@iastate.edu

Mijinyawa Yahaya and Adetunji J. A. (2005). Evaluation of Farm Transportation System in Osun and Oyo States of Nigeria .Agricultural Engineering International: the CIGR Ejournal. Vol. VII. LW 05 004.

Nelson, W. 1989. Continuous-curvature paths for autonomous vehicles. Int. Proc. Ieee international conference on robotics and automation, 3:1260-1264..

Reid, J.F. and S.W. Searcy. 1987. Vision-based guidance of an agricultural tractor. IEEE Control Systems 7(12):39-43.

Reid F. John, 1998. Precision Guidance of Agricultural Vehicles an ASAE International Meeting Presentation UILU-ENG-98-7031

Van Der Lely, C. ,1985. Tractor having guidance system. U. S. Patent No. 4515221.

Vladut Valentin, INMA Bucharest (2006). Studies and researches regarding the present stage of harvesters with an axial thresher on a world level.

CHAPTER 5

Agricultural Postharvest Systems

5.   Introduction

The products of primary production processes do not usually meet the needs of the consumers, neither has the desirable value addition to enhance its value on the international market, hence the need to further process them into more finished products which is the exclusive role of a post harvest systems engineer. Also, agricultural products are highly perishable goods and are susceptible to sudden changes in environmental condition, as such needs special handling procedure and storage condition which is also within the jurisdiction of post harvest systems engineering.

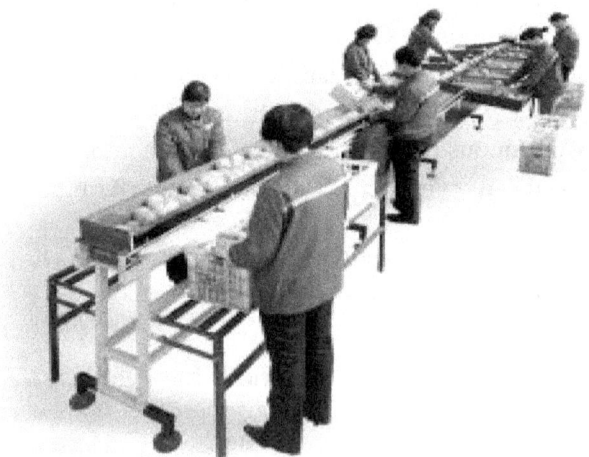

Figure 5-1: Fruit processing plant

Their therefore scope covers the processes and machines required to convert agricultural raw materials or products into finished consumer goods. It involves harvesting, transporting, handling, storage, processing and packaging. Food and processing agricultural engineers combine design expertise with manufacturing methods to develop economical and responsible processing solutions for the industry as well as look for ways to reduce waste by devising alternatives for treatment, disposal and utilization. Few among several products processing techniques are described in the following sections.

5.1 Drying of agricultural materials

Drying is the reduction of moisture content to a given final value at which the material can be stored. During drying, the moisture content of a product reaches equilibrium with the moisture content (relative humidity) of the surrounding air. In general safe storage of a product is reached below or at the equilibrium moisture content that corresponds with a relative humidity of 70% or lower. For sowing seed, the *upper limit* is 40%, for tubers the *lower limit* is 80%.

A product loses water (i.e. it dries out) when the relative humidity of the drying air is lower than the equilibrium relative humidity that corresponds with the moisture content of the product. In comparison, the more the difference between relative humidity of two particular products, the faster the drying process.

The drying process for bulk agricultural materials is in practice a very complex process owing to various disturbing phenomena such as shape of individual grains deviating from spherical and materials not being regarded as homogeneous as regards to moisture conduction. The majority of agricultural products, which are dried, may be regarded as solid, porous, or coarse material in a loose bulk state (in a layer or pile).

Drying mechanisms and systems

During drying water evaporating from the surface of the material is removed by air.

Moisture migrates to the surface under the effect of moisture gradient formed between the inner parts and the surface. This process lasts until equilibrium is attained between the inner parts and the surface, and between the surface and the ambient.

Drying process is divided into three characteristic zones in which drying vary.

Stage 1: Moisture movement

Moisture moves under the effects of capillary and osmotic forces from the inside to the surface of the material. It is known that warm air is lighter than cold air, so heated air rises by itself (vertical air current). Opposed to this is the horizontal air current of the wind. These two natural air currents are not very powerful: their speed decreases quite quickly when they meet obstacles. Evaporation zone is at the surface. Drying rate is maximum and constant at this stage.

Stage 2: Moisture drop

Moisture content drops below the maximum hygroscopic moisture content and the surface of the material dries to the equilibrium moisture content corresponding to the drying air. A faster air current through the products causes the moisture equilibrium between product and drying air to be reached sooner. Evaporation zone advances due to higher resistance and temperature increases.

Stage 3: Moisture removal

This starts when the moisture content of the material is less than the maximum hygroscopic content. Drying rates decreases and tend to zero. The hygroscopic content is below storage value, about 14%) and hence this storage has no practical interest. Drying of injured and broken kernels occurs significantly move rapidly than that of a pile consisting of a sound/whole grains. During high-temperature drying a product may undergo browning, implying a reduction of its feed value.

Figure 5-2: Frying of garri in open dryer

Drying systems

Selection Systems for drying grains range from thin layer drying in the sun or a simple maize crib to expensive mechanized systems such as continuous flow driers. The choice is governed by a number of factors including: rate of harvest, total volume to be dried, storage system, cost, and flexibility. Drying systems fall into two principle groups: Natural drying and artificial drying.

1. *Natural drying*

Natural drying is the use of ambient air temperature and either direct sunlight or natural air movement to remove moisture through the crops to a safe level for storage. The traditional methods used by farmers for drying grain rely on natural air movement to reduce moisture content to a safe level for storage. In addition they may utilize the extra drying capacity gained by exposing the produce to the sun. With good ventilation through the store the grain can be harvested just after it is ripe (about 30% MC for maize) but most methods allow some of the drying to take place naturally while the crop is still standing in the field.

Natural drying of agricultural materials occurs in three ways as follows:

a. Drying in the field before harvest
b. Drying in shallow and exposed platforms to sun and wind on a surface which prevents moisture from the ground to reach the produce.
c. Drying in or on a structure which has open sides to permit air movement through the bulk.

a. Field drying

The method of leaving the crop standing in the field for drying is popular in areas where maturity of the crop coincides with the beginning of dry season. Field drying of the crop often delays the clearing of the field for net cropping season.

Drying in a completely dry harvesting period

The product can be dried in a simple way with the help of the sun and wind:

i. Drying on the stem in the field before harvesting. It offers no protection against birds, for example. There is danger that the grains will shatter.
ii. Drying after the harvest on various kinds of racks which allow free circulation of air (Figure 5-3). The principle is that the natural current of the dry air (the wind) is hindered as little as possible on the outside. This method offers very little protection against other dangers that threaten the product (insects, birds, rodents and theft).

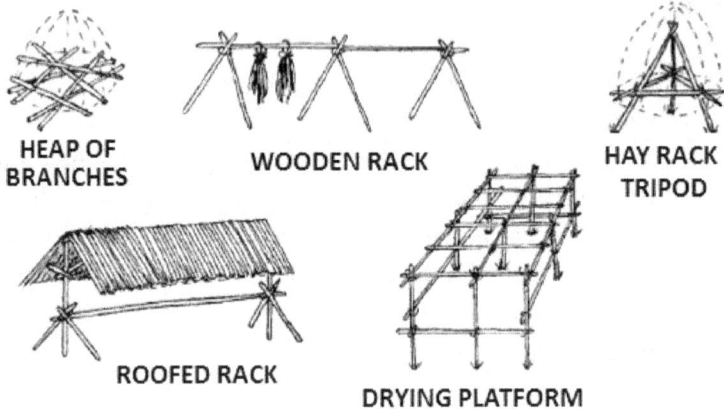

HEAP OF BRANCHES

WOODEN RACK

HAY RACK TRIPOD

ROOFED RACK

DRYING PLATFORM

Figure 5-3: Different types of racks

Drying in dry period with occasional shower and cool nights

It is very harmful for a product if it suddenly becomes wet again during drying. This will cause it to crack or burst. If the product is dried unthreshed the product should be on the inside and the foliage or straw as much as possible on the outside as extra protection against rain.

Methods employed include:

i.   Drying on roofed racks where the wind reaches the product but the rain is kept out.
ii. Drying in loosely built stacks (Figure 5-4).

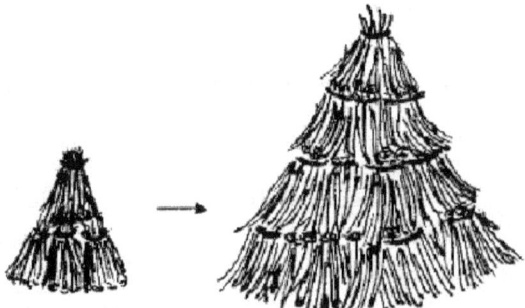

Figure 5-4: Stack of sheaved cereals

iii. The threshed product is spread out on a big piece of canvas in the sun. During rain and at night the canvas is tied or folded together, and if possible brought inside. Because the product is directly exposed to the sun, the thin layer (e.g. 3 cm) of grain should be evenly mixed every quarter of an hour or so. This method requires more attention but offers more protection against other dangers.

Drying in a wet harvesting period

Drying in wet season can be done

a. In a maize crib or in airy baskets. With this method the outside air, can blow through the unthreshed, loosely packed product as much as possible.
b. With the help of the flue gasses and the warm air of a fireplace. This can be done for example by constructing an airy platform above the fireplace (Figure 5-5) in such a way that the smoke and hot air can move easily through the product.

Figure 5-5: Drying above the fireplace

Take care that sowing seed is not heated above 40 °C and watch out for fire. The costs are low and the method protects reasonably against the other dangers, but deterioration of taste may occur.

Shallow layer natural drying

The harvested crop is spread on hard surface such as floors, on roof tops or purposely built platforms or trays. When exposed to the sun, the crop will dry fairly quickly depending on the humidity of the ambient air. Labour may be reduced considerably by placing the crop on a plastic or tarpaulin sheet for easy handling or on a platform/tray covered by for instance transparent plastic.

2. *Artificial drying*

A faster and more reliable, but more expensive way of drying is to bring artificially heated air in contact with the product to be dried. Artificial drying uses fan to move air through the crop with the air either at ambient temperature or artificially heated.

In some areas storage of crop is restricted to the amount which can be dried by the heat supply similar to that available from a kitchen fire.

Types of artificial driers

Large scale system driers can be derived into the following categories: barrel driers, continuous flow driers, batch driers and sack driers. They may also be either high temperature or low temperature systems. Different forms of artificial drying are characterized by the depth or thickness of grain being dried. Examples of artificial drying include: In-sack driers, barrel driers, shallow layer driers and deep layer driers.

a. In-sack drying

Grain in-sacks can be dried in a stack or the sacks may be laid one or two layers on a platform drier. A platform drier consists of a chamber with an open top of wire mesh, bamboo or other means to support 2 to 3 layers of sacks. Gaps between sacks in both platform driers and sack-stack should be filled with empty bags or straw to minimize air leakage. Secondly, as pointed out earlier grain should be cooled before being left for storage.

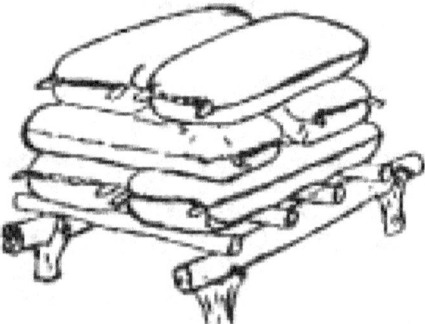

Figure 5-6: Pile drying in sacks

It is also possible to build a solar drier which can work on solar heat for most of the time, but which can, if necessary, be artificially heated during periods of heavy clouding or rain. Solar driers have the advantage of no fuel costs and can also be used for other crops such as okra, cassava, fruits and vegetables.

Disadvantages of such system include:

Temperatures could rise up to 65 - 80 °C: rice and sowing seed can be damaged and most useful only at certain hours of the day and of limited use during long periods of rainfall or very cloudy weather.

b. Barrel driers

Two designs of barrel drier is available using oil drums in which a fire is made): the simple oil drum drier and the pit barrel drier (Figure 5-7 and 5-8).

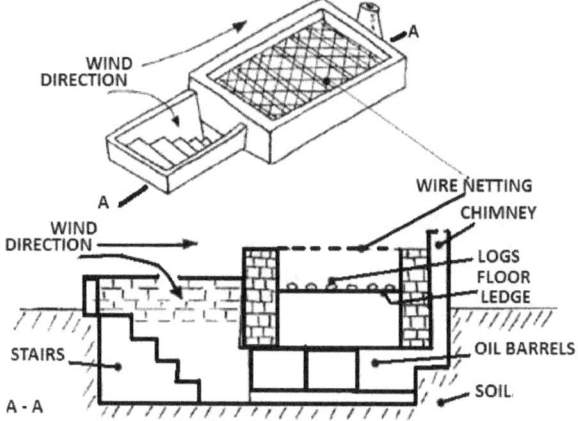

Figure 5-7: Barrel and hand-rammed earth or mudblocks driers

i. Pit barrel drier

The pit barrel drier is also known as the bush drier or brooks fire drier. By making a fire in the connected oil drums, the surrounding air is heated and rises through the product to be dried, which is spread in a not too thick layer on a screen, supported by logs.

ii. Simple drum barrel drier

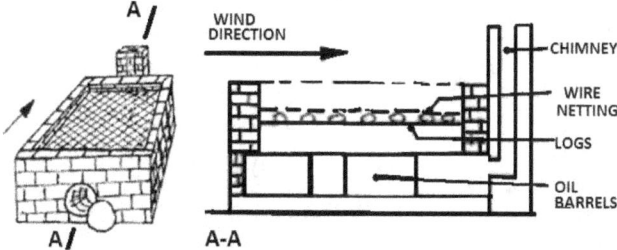

Figure 5-8: Barrel driers made out of barrels and hand-rammed earth or mudblocks

c. Shallow layer artificial driers (batch driers)

These are shallow layer dryers, often in the form of a tray with a perforated base. The dimensions may be 1 to 2m wide and 2 to 4m long with the grain bed being 150 to 300mm deep. Warmed air is blown into the plenum chamber beneath and then up through the grain. This type of dryer is suited to a smaller operation than continuous-flow driers. They may be both mechanically or manually loaded and unloaded.

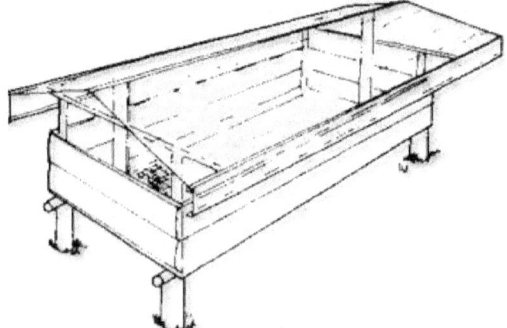

Figure 5-9: Shallow layer dryer

d. *Deep layer driers*

These consist of beds, bins, silos or rectangular warehouses equipped with ducting or false floor through which air is distributed and blown through the grain. The depth of the grain layer may be from 30cm and up to 350cm. The crop is piled over the lateral ducts which are fed with air from a main duct. The lateral ducts can be installed above or below floor level.

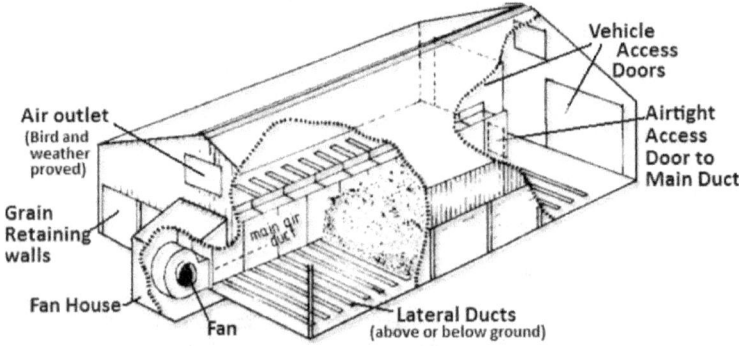

Figure 5-10: Deep layers drying

e. *Continuous-flow driers*

The grain passes through the drier in a continuous flow at a controlled rate. The grain is kept in a thin sheet, approximately 100 to 150mm deep and hot air is blown through the crop. Continuous-flow driers are high in cost and are applicable only in highly mechanized situations.

Figure 5-11: Conveyor drier

f. Solar driers

The basic mechanism of material drying is one of heat and moisture transfer between the material and the air. The heat is transferred to the surface of material by conduction and convection from adjacent air at temperature above that of the material being dried. If the air is passed through the material at a relative humidity of less than moisture content in material, the air will absorb moisture from the material while increasing its absolute and relative humidity.

Figure 5-12: A solar fruit dryer

The use of solar dryers in the agriculture to conserve vegetables, fruits and other crops has shown to be a practical, economical and responsible approach environmentally. Solar heating systems to dry food and other crops improves the quality of the product, while reducing wasted produce and traditional fuels - thus improving the quality of life.

g. *Rotary driers*

A motor driven ventilator blows either heated cooling air of the motor or air heated by a burner through the product, which is spread on a drying platform. Motor driers are suitable for drying cereals, pulses and oil containing products. The drying capacity depends on the size of the drier and on the product to be dried. Drying machinery for efficient, safe drying and post harvest protection of non granular crops, forage, wood, fruit, roots, spices etc

VENTILATOR   BURNER   PRODUCT

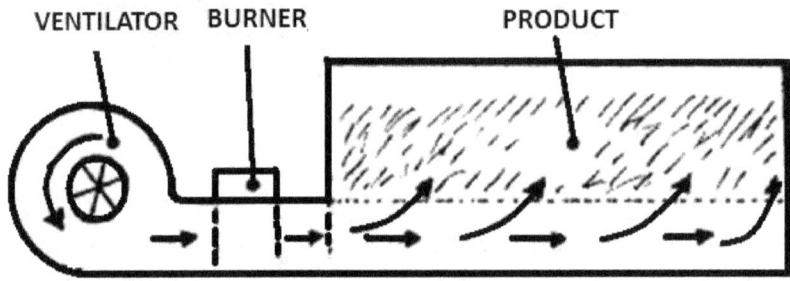

Figure 5-13: Motor drier.

Drying problems

Problems often encountered in drying include;

1. *Overdrying*: Major drying problem is that of overdrying of grains. Overdrying of grain at excessive temperature can
   a. Set up stresses in the individual kernels leading to cracking and loss of viability.
   b. Lead to loss of all moisture below the safe storage moisture content thereby leading to a loss in the value of the crop.
2. *Presence of dirt in material*: Dirty crops such as grain with a large proportion of chaff, fine seeds and dirt becomes more difficult to dry as the resistance to airflow increases due to spaces between grains being blocked.

5.2   Cooling of agricultural material

The methods adapted to cool grain after drying are dependent on the drying system used. Sun-dried grain can reach high temperatures while in the direct sunlight. If it is to be stored in any container through which air cannot freely pass, it should at least be left shaded for an hour or more before storing. Fan ventilated batch driers of all types, including sack driers, should have the fan left running with no added heat until the crop is at ambient temperature before discharging the crop from the drier.

Heat production in biological materials during storage

Agricultural products continue to live and respire during storage. The rates of these biological processes are functions of the moisture content and temperature, in addition to the structure of the material. Both biological processes and the life functions of microorganisms produce oxidation and heat is librated. The rate of heat generation is of decisive importance from the point of view of storage.

A pile containing damaged grains has a higher oxidation rate and deterioration activities even when subjected to the same external conditions. The amount of heat generated depends on when (i.e. in which stage of the ripening process) a product is harvested and put into storage. It has been observed that the rate of heat generation is higher in the first few days than during the subsequent period.

Buffer cooling storage

Failure to cool grain that has just been dried with heat may cause an increase in moisture content great enough to seriously shorten its storage life. If hot grain is allowed to cool naturally the Relative Humidity (RH) of the air in the bin will rise and, if the saturation temperature is reached or passed, condensation can cause the grain moisture content to rise again. To prevent this possibility, after drying, the grain should be cooled until ambient temperature is reached.

Low volume ventilation (LVV) or aeration may be employed to cool grain that has been put in storage. Although it can be used in conjunction with other driers as a cooling system, the main objective of LVV is to cool the grain positively at harvest time and thus prevent infestations of insects and mites and the development of mould. It must be stressed the LVV is not a drying system. Consequently if grain is too wet at the start (over 18%) it will be unlikely to store well, and for human consumption it would be preferable to start with a mc lower than 18%.

Grain cooling equipment

Fans to be used for grain cooling can either be centrifugal or single-stage axial fans. Motors ranging from 370 to 746W cover the vast majority of fan size used. They are usually small enough to be picked up by hand and run on 13 amp switched outlets. The volume of air delivered varies with the climate but should at least be 10m^3/h and tonne.

Heating and cooling of deep piles

During the storage of agricultural products (cereals, potatoes, etc) it is necessary to remove heat generated by biological reactions i.e. re-cools the product. The whole volume of a pile is not heated or cooled uniformly during these operations; rather, a heating or cooling front passes through the pile.

Moisture exchange with air in agricultural products

For high-quality storage of crops two basic requirements must be met: Uniform maintenance of the optimal temperature, and the preservation of moisture content. Moisture content of most frosh agricultural products such as fruits and vegetable is usually between 80-85%, with a loss of few percent of the moisture content, the tugor pressure decreases and the product withers. Thus the preservation of moisture content is important in maintaining quality. The natural removal of water during storage is a result of slow diffusion processes. Dalton's law describes the exchange of moisture between fruits or vegetables and the air as expressed in the following relations.

$$G = \beta\gamma F(C_s' - C_s\varphi) \dots\dots\dots\dots\dots\dots\dots\dots\dots .5.1$$

Where
    C' = Concentration of saturated water vapor on the surface
    C_s = Concentration of saturated water vapor in the air
    φ = Relative air humidity.
    γ = Moisture exchange coefficient of the material.
    β = Evaporation coefficient

Moisture content of products grown on irrigated fields is generally higher, but they also discharge water more easily and so their γ (gamma) values are higher. These products require more careful storage, and can be preserved. Moisture content of products grown on irrigated fields is generally higher, but they also discharge water more easily and so their γ (gamma) values are higher. These products require more careful storage, and can be preserved.

5.3 Densification of agricultural materials

Certain materials (forage materials for instance) must be compressed during agricultural technological processes in order to reduce their volume to obtain a definite shape which facilitates handling or to obtain juice (e.g. grape pressing). The general relationships for pressing may be studied most simply by means of a pressing cylinder stuffed with a quantity of material and compressed slowly by a piston to decrease its volume.

The pressure exerted by the piston increases progressively with increase in volumetric weigh (density). The compressibility of forage materials depends mainly on the plant species and moisture content.

Densification of agricultural materials

Densification is the use of mechanical pressure to reduce the volume of agricultural matter and the conversion of this material to a solid form, which is easier to handle and store than the original material. Densification of agricultural residues may be used as fuel for the generation of energy.

Many researchers have tried the densification mechanism of different biomasses. It involves application of attractive forces between solid particles, interfacial forces, capillary pressure, adhesive and cohesive forces, mechanical interlocking behaviour and formation of solid bridges. Strength of the compactness depends upon material characteristics and process variables like pressure, temperature and use of binding materials.

A comparative study has been made for different methods. Mechanics of agricultural materials, as a scientific discipline, presently is being developed and so far there are many process-material interactions that do not have the same methods of representation. Nevertheless, the experimental methods that have so far been developed can somehow be used successfully to select, design and optimize such process machines (Sitkei, 1986).

Figure 5-14: An extruder (Munoz-Hernandez *et al*, 2004)

Some of the methods available for compaction of the residues are the piston press, screw press, roller press and palletizing machines. This section reviews the different biomass densification methods and its mechanism.

Requirements for densification

Requirements for the design, construction and improvement of densification systems are based mainly on the knowledge of suitable levels of process variables and material variables.

a. *Process variables* include die geometry, relaxation time, die and material temperature and pressure.
b. *Material variables* include moisture content and moisture distribution, size and shape of particles; particles size distribution, biochemical and mechanical characteristics etc.

These variables are adjusted to achieve the highest density, the biggest output, the best consistency (durability) and the lowest power consumption. (Munoz-Hernandez *et al.*, 2004)

Methods of densification

Four methods of achieving densification using commercial machines include: baling, cubing, pelleting, and briquetting. These processes can be achieved by means of piston presses, extrusion screws or by roll presses.

Figure 5-15: Products of densification processes

Cubing, baling and pelleting processes have been frequently used in animal feed production while briquetting by means of piston presses and screw extruders have been used in solid fuel manufacture. The roll press has been used mainly for metallic and mineral dust compaction.

Pelleting

A basic method for reducing the volume of forage and granular-farinaceous fodders in material such that separation of the individual component is prevented is referred to as pelleting. Pelleting is capital- and energy-intensive feed manufacturing operation. It is a key driver of feed mill profitability. Pelleting of feed also provides the benefits of:

1. Increasing the bulk density of feed;
2. Improving feed flowability; and
3. Providing opportunities to reduce feed formula costs through the use of alternative feed ingredients.

The energy requirements of pressing by pelleting machines comprise of the net pressing work and the work spent in pushing. The net pressing work may be determined most simply by means of a pressing cylinder while the area under the pressure curve gives the work spent in pressing i.e.

$$A_c = F \int p \, ds \, ... 5.2$$

Fodder flours are pelleted mostly in die rings. The farinaceous material charged is compressed by a suitably adjusted roller and pressed into the boreholes (channels) of the ring. A knife on the outer side cuts the compressed material emerging from the channels. In so a compressed layer (carpet) is formed on the running surface of the ring.

The layer thickness of fresh material charged before the roller is gradually reduced as a result of compression by the latter. With increasing pressure the thickness of the layer decreases gradually until it reaches the minimum value. The pressure attains its maximum value at a point midway between the work area and the roller presses the materials into the extrusion channels.

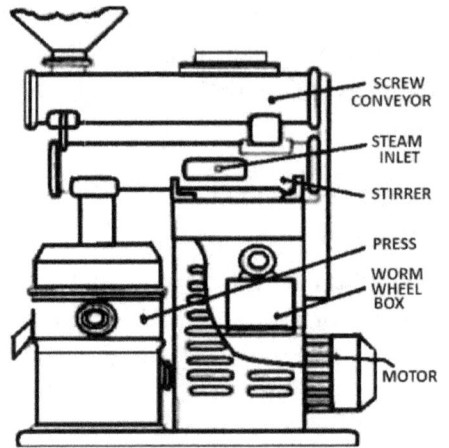

Figure 5-16: Powder feeder seed pelleting equipment

The pressure remains practically constant in the material during pushing and then decreases rapidly as the gap increases. The pressure required for pelleting varies between 500 and 1500 bar. The diameter of fodder flour pellets is 5-15mm and of forage fodder pellets 20-50mm.

Terms used in pelleting

Roller assembly: This is simply a cylinder idling on bearings in much the same manner as the front wheel of a bicycle. The only driving force acting on the roller assembly is the frictional turning force from the die, acting through a very thin layer of feedstock between the die and the roll.

Die: The die is the rotating, driven component, utilizing the power applied to the pellet mill. The die is composed of a ring of steel perforated with holes through which material flows at pellet density. Perforation diameter and die thickness determine the final pellet size and quality.

Feed: This is the material to be pelleted after it has been properly conditioned for extrusion.

Work area: The work area in the pelleting chamber can be defined as that area where the material is received at its own density, compressed and then forced through the holes in the die. In reality there are two supporting sections of the work area.

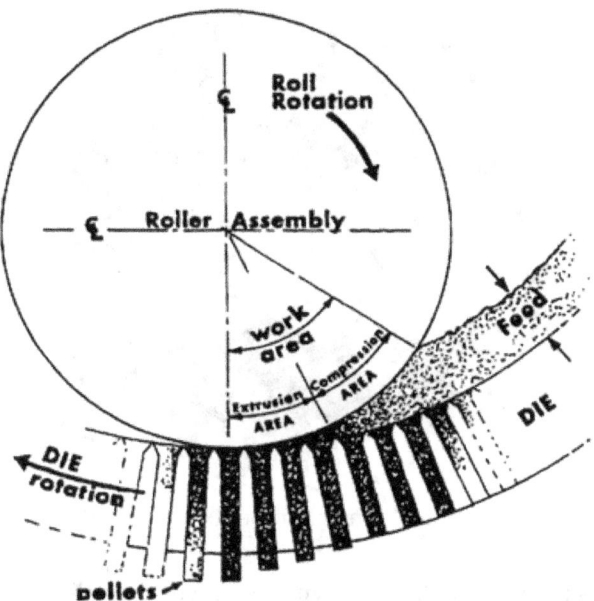
Figure 5-17: Roller die configuration

Compression area: Here the feed is compressed to near pellet density, forcing out the entrained air, and forcing the individual particles into intimate contact with each other.

Extrusion area: Here the feed has reached pellet density and is forced to flow through the die openings.

Pelleting process and mechanism

A pelleting mechanism (Figure 5-18) undergoes the following processes in material handling:

1. Incoming feed flows into the feeder is delivered uniformly into the conditioner for controlled addition of steam and/or molasses.
2. From the conditioner, the feed is discharged over a permanent magnet into a feed spout leading to the pelleting die (1).
3. Inter-elevator flights in the die cover feed the mash evenly to each of the two rollers. (2)
4. Feed distributor flights (3) distribute the material across the face of the die.
5. Friction driven rollers (2) force the feed through openings in the die as the die revolves.

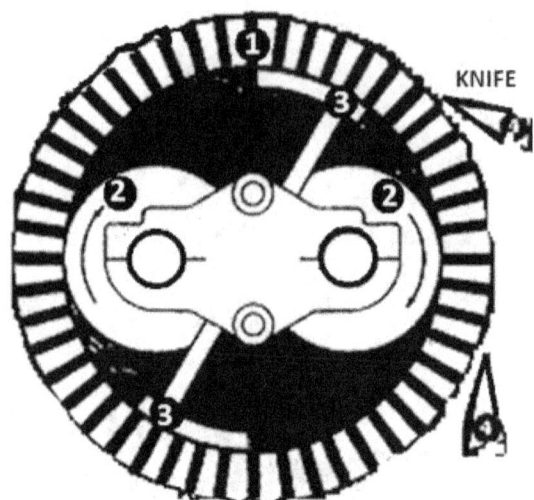

Figure 5-18: The pelleting chamber of a two roll pellet mill

6. Cut-off knives (4) mounted on the swing cover cut the pellets as they are extruded from the die.
7. The pellets fall through the discharge opening in the swing door.

Briquetting of agricultural materials

Biomass briquetting process is known as high compaction technology or binder less technology in which biomass residues are compressed under high temperature and pressure. The quality of the product/compact depends on the pressing technology, characteristics of the pressed material and on real conditions during pressing. The compressed materials form small logs with a diameter of between 50 and 100mm and of any length depending on the briquette technology used; screw, piston, and oil-pressure.

Briquettes are commonly made in piston or screw presses, which produce briquettes with uniform density and stability and the utilization of these machines is gradually increasing especially in European countries and China (Zeng *et al.*, 2007). The briquetting processes has developed in two directions: firstly, Europe and the USA have chosen the path of mechanical compression (hydraulic or pistons), secondly, the East has preferred worm screw pressing.

The piston press and the screw press are common technologies for biomass briquetting in India and Bangladesh (Bhattacharya and Kurma, 2005). In briquetting processes compression takes place under 200 °C to 250 °C inside taper die and produces denser and stronger briquettes than those produced by piston presses (Sharif *et al.*, 2008).

Figure 5-19: Biomass briquetting system (Bhattacharya and Kurma, 2005)

Heated-die screw press briquetting is a popular densification method suitable for small-scale operations in developing countries. In this method, the raw material from the hopper is conveyed and compressed by a screw that forces it through a heated die.

This process can produce denser and stronger briquettes compared with piston presses. The briquetting machines employed in the technology packages of RETs in Asia programme are of the single extrusion heated-die screw-press type.

Dewatering of agricultural materials

The extraction of juice or water from whole crop mechanically is obtained by pressing. Pressing has recently also been utilized for dewatering green fodders and preparing protein concentrates. Energy requirements for mechanical dewatering are about 100 times less than those of thermal dewatering. Energy requirement of dewatering depend on the extent of liquid removal.

Similar methods are employed in fruit juice manufacturing industry and in other products. When pressure is applied to a plant body, the thin walls of the fruit cell breaks upon and the liquid contained in the later migrates toward the free surface in the direction of the pressure drop.

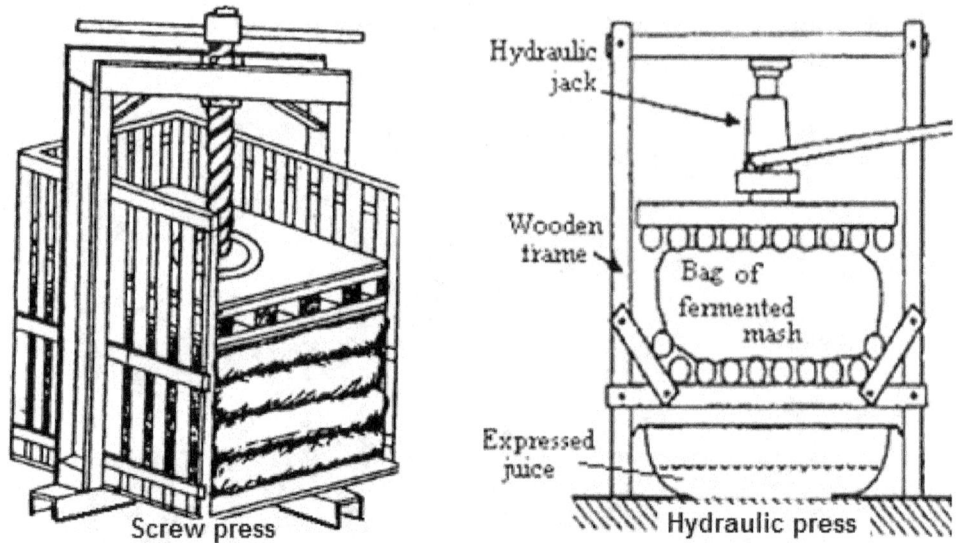

Figure 5-20: Cassava mash presses (CIGR 1999)

Dewatering is greatest during this first stage. After the removal of a part of the juice from the material, the solid parts support the pressure increasingly, where by the pressure in the intermediate pores decreases and juice leaves the material by seepage at a decreasing velocity.

Figure 5-21: A hydraulic press for honey extraction

Dewatering takes place over a period of time and the period for which pressure is maintained plays an important role in determining the quantity of juice extracted. The dry matter content of the juice extracted depends on numerous factors such as

1. The pressure and temperature,
2. The duration of pressing,

3. The thickness of the layer pressed,
4. The extent to which the material is destroyed etc.

Juice pressed from agricultural materials contains both dissolved and solid dry material and therefore the dry content of the initial material also decreases during pressing.

5.4   Cutting of agricultural material

Cutting is used frequently during harvesting of agricultural material, separation and subsequent comminution of plant components. Cutting is also the main operation in fodder preparation. During cutting, a cutting edge (knife) penetrates into the material, overcoming its strength and cutting thereby separating it.

The cutting process of forage crops is greatly influenced by its physical and rheological properties. Beyond the physiological maturity stage, moisture content of forage crops such as sorghum and maize decreased sharply as the age of the plant increased, whereas, the stem diameter decreased with the ageing of the plants (Jekendra, 1999).

Without knowing the optimum cutting energy requirements of forage crops, it is hardly possible to design an efficient forage harvesting machine. The shear strength of maize and sorghum stems increased with the decrease in moisture content. Beyond the physiological maturity stage, further increase in the plant age decreased the stem diameter.

Cutting methods

Four basic methods of cutting employed in agricultural operations as identified include:

1. *Counter moving blades*: Two sets of blades participate in this cutting (Figure 5-22). The knives move in opposite direction with the material in-between the moving blades.

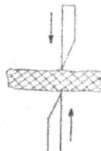

Figure 5-22: Counter moving blades

2. *Cutting by means of a resting and moving blade*: The resting blade supports the material while the moving blade slices the material against the stationary blade (Figure 5-23).

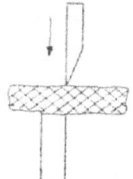

Figure 5-23: Methods of cutting

3. *Cutting of thin layers*: The stress distribution around the cutting blade (edge) is significantly distorted by the free surface found close to the cutting plane (Figure 5-24). The material may be fixed rigid e.g. beet cutting.

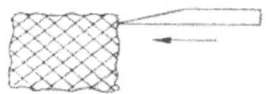

Figure 5-24: Cutting of thin layers

4. *Free cutting*: One end of a relatively long stalk is fixed and counter support is ensured by the moment of inertia of the stalk (Figure 5-25). The velocity of cutting edge is high; 20-40m/s.

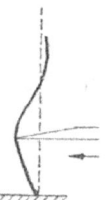

Figure 5-25: Free cutting

Products may be cut individually or in bundles depending on the type of material and the technological process. Material is first compressed and deformed under a cutting edge. The cutting edge may move normally to the material or at a certain angle. In the later case, the cutting edge will be displaced during cutting in the direction of the surface cut.

Energy requirements in cutting

Two stages are distinguished in cutting process:

1. *Preliminary compaction* of material until a pressure is reached at which the material under the cutting edge yields

2. *The motion of the edge into the material*: The specific energy requirements of cutting may be obtained by dividing the total work by the cross-sectional area of cutting.

$$F = \frac{A}{A_f} \ldots\ldots\ldots\ldots\ldots\ldots\ldots\ldots\ldots .5.3$$

Where

F = Specific energy requirements

A=Total work

A_f = Cross sectional area of cutting

Cutting is not a static but a dynamic process. Increasing cutting velocity decreases initial compaction as a result of the material's inertia and plastic behaviour whereby energy requirements are lowered. Such factors as age of material, thickness of the cutting edge and the angle of sharpening (bevel angle) influence the cutting resistance and energy consumption.

Determination of cutting forces

Penetration of the knife into the material first causes compaction up to a height h, until a given pressure is reached at which rupture occurs. At this point, which is at the instant of cutting, a force, P appears on the cutting edge (Figure 5-26). In advancing, the knife compact successive regions of materials until rupture occurs again.

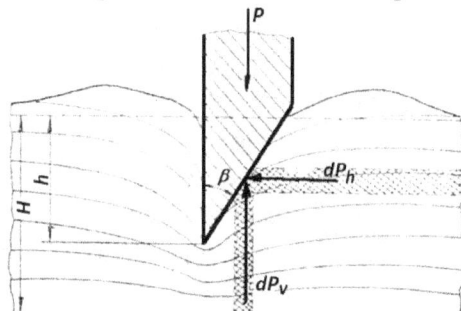

Figure 5-26: Kinematic relations of materials under cutting knife (Sitkei, 1986)

The distance between individual ruptures is a function of the cutting thickness and the angle of the edge. Two forces acts on the knife, one on the sharpened face and the other at the vertical side of the knife. The normal force acting on the sharpened face decomposed into horizontal and vertical components as follows.

$$N = P_x Sin\,\beta + P_h Cos\,\beta \ldots\ldots\ldots\ldots\ldots\ldots\ldots .5.4$$

Free cutting

Rotary grass movers contain no counter blade, so the process is one of free cutting. Since reaction force corresponding to the maximum cutting force arises in the material. The components of this reaction force include the mass inertia resistance to acceleration) of the stalk Being cut and the static reaction forces due to banding of the stalk and angular displacement of the cross-sectional area cut.

$$P = P_1 + P_2 = N\,Cos\lambda + T\,Sin\lambda \quad ....5.5$$

The force N acting normally to the cutting edge is the product of the specific cutting resistance per centimeter and the cutting length $N=Pl$. while the tangential force, T is the product of the friction coefficient and the force N i.e. $T= \mu N$.

Using these relations, the peripheral force is written thus.

$$P = \rho l(Cos\lambda + \mu\,Sin\lambda) \quad .... 5.6$$

The resultant force R and the displacement vector μ do not coincide in a straight line $(\mu \neq \rho)$

5.5   Size reduction of agricultural materials

Size reduction of granular products and forage fodder is an important operation in agricultural material processing and preparation. Some grains will reduce more easily than others because of their structure and composition. Routine evaluation of ground grain or feed particle size will assist the on farm feed processor to better manage the grinding process. Particle size reduction increases the surface area of agricultural materials like grains, thus allowing for greater interaction with digestive enzymes. It also improves the ease of handling and mixing characteristics (Goodband *et al*, 2002).

Size reduction methods

There are four main size reduction methods identified and is in common use: impacting, grinding, crushing and sawing (Figure 15-27). Impacting is suitable for hard and brittle raw materials, such as maize feed; sawing is better for large and fragile feed; and crushing and grinding are used for tough feeds.

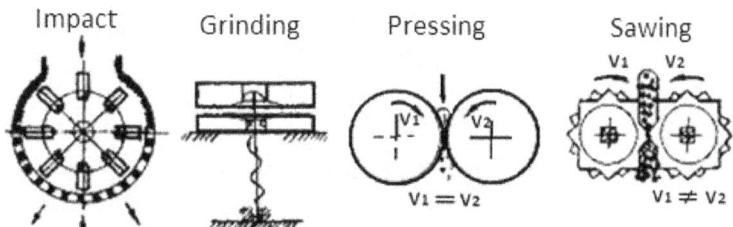

Figure 5-27: Grinding methods for feed

Each of these methods employs one or more of four basic force actions: impact force, attrition force, shear force and compression force.

a. *Attrition action:* Attrition is a term applied to the reduction of materials by scrubbing it between two hard surfaces. Grinding on mill-stone is a typical attrition process. Hammer mills equally reduce by *grain size by* attrition combined with shear and impact with close clearances between the hammers and the screen bars.

Figure 5-28: Size reduction by attrition

Though attrition consumes more power and exacts heavier wear on hammers and screen bars, it is practical for crushing the less abrasive materials such as pure limestone and coal.

b. *Shear action:* Shear consists of a trimming or cleaving action rather than the rubbing action associated with attrition. Shear is usually combined with other methods. For example, single-roll crushers employ shear together with impact and compression

Figure 5-29: Material shearing

c. *Impact action:* Impact refers to the sharp, instantaneous collision of one moving object against another. Both objects may be moving, such as a baseball bat connecting with a fast ball, or one object may be motionless, such as a rock being struck by hammer blows.

Figure 5-30: Impact grinding

d. *Compression action:* Material reduction by compression is done between two surfaces, with the work being done by one or both surfaces. Compression could result in milling action or crushing action.

Figure 5-31: Size reduction by compression

1. *Size reduction process by impact action*

Size reduction by impact can be by gravity or dynamic action. Coal dropped onto a hard surface such as a steel plate is an example of gravity impact. Gravity impact is most often used when it is necessary to separate two materials which have relatively different friability. Material dropping in front of a moving hammer (both objects in motion), illustrates dynamic impact.

When crushed by gravity impact, the free-falling material is momentarily stopped by the stationary object. But when crushed by dynamic impact, the material is unsupported and the force of impact accelerates movement of the reduced particles toward breaker blocks and/or other hammers.

Impact equipment

The most common pieces of equipment used to reduce the particle size of grains by impact are the hammer mills. There are different types of hammer mill, the choice of which to use depends on the unique requirements of every individual situation.

Hammer mill

Hammer mills are among the oldest, yet the most widely used size reduction machines in feed mills. Although recent years have witnessed the introduction of new types of hammer mills, many of them are a refinement of the basic hammer mill designed in order to serve more specialized purposes. Hammer mills crush materials in two stages:

Stage 1: The material is reduced by dynamic impact;

Stage 2: Crushing then occurs by attrition and shear in the second zone, where small clearances exist between hammers and screen bars. This second zone is the final sizing zone for the product. Hammer mills have high reduction ratios and will produce high capacities whether used for primary, secondary or tertiary crushing.

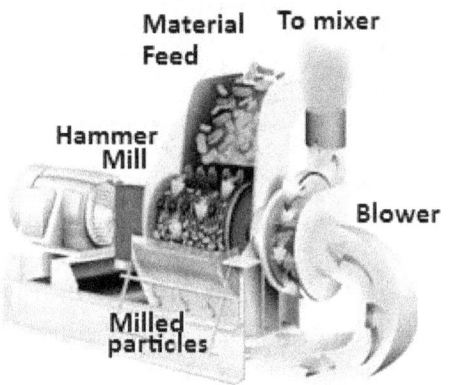

Figure 5-32: Material flow inside a hammer mill

Hammer mills can be divided into two generic types on the basis of orientation of the rotor; namely, horizontal and vertical. Both types have hammers that rotate within the shredder and cause particle size reduction through collision with the infeed material.

1. Horizontal swing hammer mill

The horizontal swing hammermill is commonly used in mixed waste processing. Its principal parts are the rotor, hammers, grates, frame, and flywheel. Its rotor and flywheel are mounted through bearings to the frame. A set of grate bars or cages through which size-reduced materials exit the machine is held in the bottom of the frame. A diagrammatic sketch of a horizontal hammermill is shown in Figure 5.

In a horizontal hammermill, designed for mixed waste processing applications, the rotational speed of the rotor is usually in the range of 1,000 to 1,500 rpm. Objects to be size reduced are introduced into the infeed opening of the machine. They then interact with the hammers and each other until at least one of their dimensions reaches a size small enough for the particle to fall through the grates at the bottom of the machine.

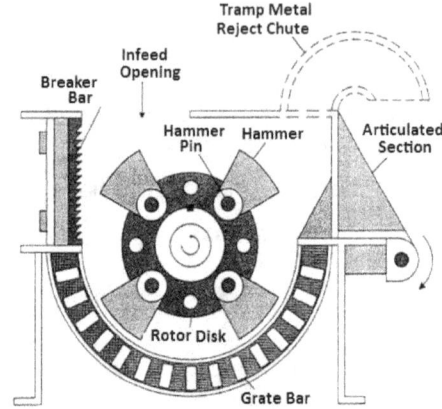

Figure 5-33: Cross-section of a horizontal hammer mill

Residence time of the material in the mill and the size distribution of the size-reduced product are largely determined by grate spacing. Other factors that affect product size distribution are feed rate, moisture content, and hammer speed (i.e., velocity of the tip of the hammer).

2. *Vertical rotor hammer mill*

A diagrammatic sketch of the vertical type of hammer mill is presented in Figure 5-34. As is indicated in figure, the axis of the rotor is in the vertical position. The in-feed material drops parallel to the shaft axis and is exposed to the action of the rotating hammers.

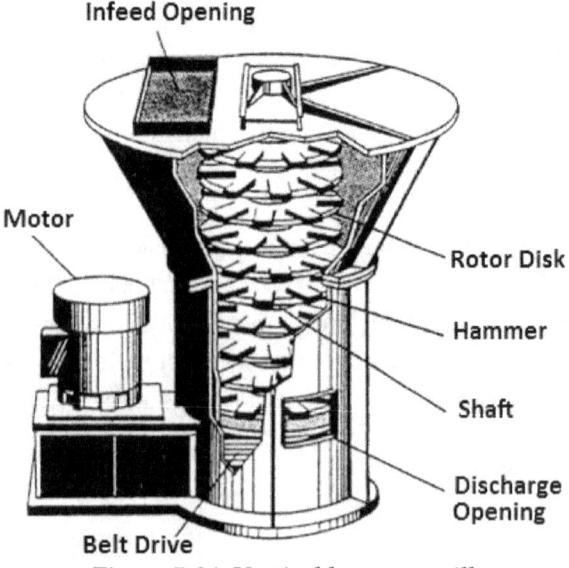

Figure 5-34: Vertical hammer mill

It is shredded by the time it is discharged at the bottom of the machine. A photo of a commercial vertical hammer mill, including the infeed hopper, is shown in Figure 3-34, with the hammers exposed for maintenance.

Advantages of hammer mill

Advantages of hammer mill over roller mill include

1. Production of a wide range of particle sizes,
2. Works with any friable material and fiber,
3. Less initial purchase cost compare to roller mill,
4. Offer minimal expenses for maintenance, and generally features uncomplicated operations.
5. The hammer mill is well suited for straw feed.

Disadvantages of hammer mill

Disadvantages includes

1. It provides less efficient use of energy compared to roller mill,
2. May generate heat,
3. May create noise and dust pollution, and
4. Produces greater particle size variability (less uniformity).

Types of hammer mill

Hammer mills are either tangential-feed or axial-feed types, according to their structure. The mill comprises a feeding part, a grinding chamber and a collector. The feeding part comprises a feed hopper and a feed control flap. The grinding chamber consists of a rotary disk, a hammer, a serrated plate and a screen. The major parts of the collector include a fan, a feed conveying tube and a collection hopper.

Axial feed hammer mill

Axial feed mill differs from the tangential-feed mill in both the direction of feeding and in primary cutting action. Straw fed from the axial-feed hopper is firstly chopped into small pieces by the primary cutting mechanism fixed in front of the grinding chamber, and these pieces then fall into the grinding chamber.

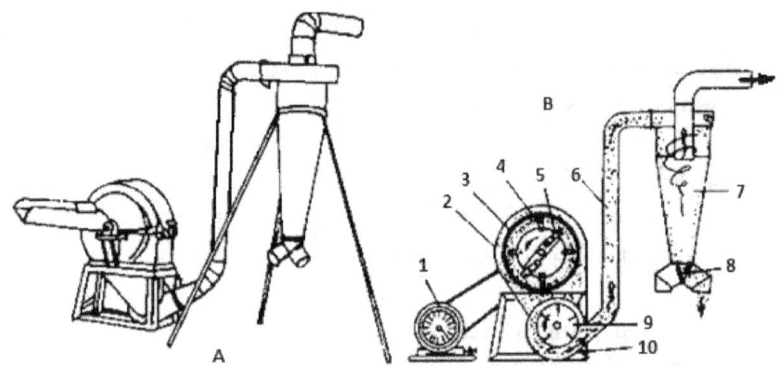

Figure 5-35: Axial feed mill (A. Outline; B. Operation)

[1. Motor; 2. Housing; 3. Ring screen; 4. Hammer; 5. Primary cutter; 6. Conveying tube; 7. Feed collection (gathering) hopper; 8. Discharging door and tube; 9. Fan; 10. Frame.]

Tangential feed hammer mill

Fed from the feed hopper in a tangential direction, the material is impacted and driven to the grinding chamber by the rotating hammers with high speed. The material in the grinding chamber is first hit and ground to some extent by hammers, and then thrown at high speed at the serrated plate and the peripheral screen fixed inside the chamber to be further ground through impact with the serrated plate and friction with the screen.

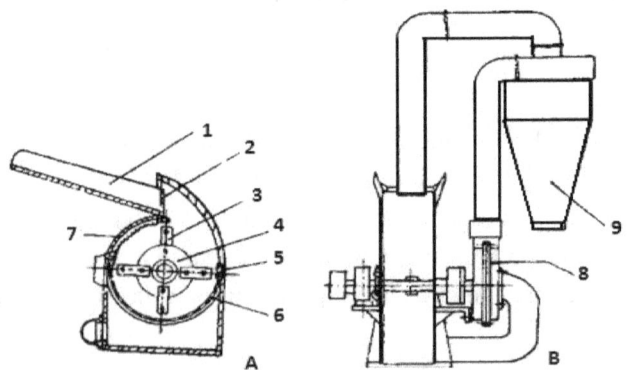

Figure 5-36: Tangential-feed grinder

[1. Feed hopper; 2. Feed control; 3. Swinging hammer; 4. Rotary disk; 5. Small serrated plate; 6. Peripheral screen; 7. Large serrated plate; 8. Fan; 9. Feed gathering (collection) hopper.]

Thus the load on the grinding chamber is reduced; and the feeding capability and efficiency of processing improved. The axial-feed mill is especially suited to grinding straw with high moisture content.

The reduction of grains into small particles using a hammer mill is determined by the number of hammers, hammer size, hammer wear, hammer tip speed, motor horse power, screen area, diameter of screen opening, grinding rate i.e. Rate at which grain is fed into the grinder) and grain quality (moisture content, hardness, grain test weight / bulk density, grain type etc (Herrman and Harner, 1995).

Composite feed mill

This is basically a hammer mill type machine consisting of an intake hopper, the milling portion, a screw conveyor and a horizontal feed mixer so that the combination serves as a semi-automatic feed mill. Many models and sizes are available in the market. The mill requires and additional screw conveyer for sending the products from the hammer mill to the feed mixer and a storage mill above the mixer.

Technical considerations for choice of mill

Due to high size reduction ratio, good control of particle size range with relatively good cubic shape of particles, hammer mills are widely used (Nikolov 2004, Mani *et al.*, 2004). For many mash (meal or non-pelleted) feed, hammer mills have been traditionally used to produce finer ground feed components.

Hammer mills have a power requirement that ranges between two and 50 kW, while motor-driven plate mills generally demand less power and 0.5–12 kW is sufficient. As a rule of thumb, about 1 kW can mill 25–30 kg of produce per hour (Clarke and Rottger, 2006). Machines and spares for hammer mills are often made locally at lower prices than imported parts. Some components, such as screens, are rarely made locally and need to be imported, especially those of a very fine size.

Knife mills work successfully for shedding forage under various crops and machine conditions (Ige and Finner, 1976). Due to increasing energy costs, increasing awareness of feed quality and emerging environmental concerns, the validity of the hammermill as the only alternative for particle size reduction (grinding) applications is being challenged by the use of roller mills.

Plate mills of 0.5 kW are usually made for grinding soft fruits and vegetables, and the plates can be made from locally produced steel. Plate mills are not popular for grain milling as they have to be manufactured from chilled cast iron, which is rarely available in Africa.

2. *Size reduction by grinding (comminution)*

The aim of grinding/comminution is to reduce the size and increase the specific surface area of particles. Material breaks when the local load transferred by impact exceeds the breaking strength of the material and the energy transmitted is sufficient to overcome cohesive forces at the new surfaces created.

In milling industry grinding is done in most cases by hammer mills; wheat is ground by cylindrical comminuting equipment (roller mills). The following fractions are typical grain (particle) size distributions used in fodders.

a. *Grit*: Fine grit = 90% of grain diameter less than 1.10mm
     Medium fine grit = 90% of grain diameter between 1.0-2.0m
     Coarse grit = 90% of grain between 3and 5mm

b. *Middling*: 0.12-0.3mm grain diameter
c. *Flour*: 0.07-0.2mm grain diameter

Fineness of grind

As long as feed ingredients have been processed, determining and expressing fineness of grind has been a subject of study. While the visual appearance or "feel" of the product may allow an operator to effectively control the process, subjective evaluation is inaccurate at best. Descriptive terms such as coarse, medium, and fine are simply not objective. What is "fine" in one mill may well be "coarse" in another.

In terms of finished particle size(s) produced, describing the process or equipment is also subject to wide variation in terminology. The condition of the corrugations (roller mills), condition of the hammers and/or screens (hammer mill) as well as other factors such as moisture content of the grain can and will produce widely different results.

The quality of the grain or other materials being process will also have a profound impact on the fineness and quality of the finished ground products. A sieve analysis is the best and most descriptive way of measuring the finished particle size. The results are expressed in terms of mean particle size; or percentage (ranges) on or passing through various sized sieves.

Excessive levels of fine or coarse particles, etc. will be indicated as well as the average particle size. Distribution of the various micron sizes will also be indicated.

Typical descriptions of an objective measurement might be: corn ground to 750 Microns, not less than 50% passing through a 20 mesh, and not more than 85% through a 40 mesh.

3. Size reduction by pressing

This accomplishes size reduction through a combination of forces and some design features. Material is sandwiched between two surfaces under the action of compression or friction rolling. Jaw crushers for instance use compression/pressing method (material being sandwiched between two surfaces) to reduce extremely hard and abrasive rocks. However, some jaw crushers employ attrition as well as compression and are not as suitable for abrasive rock since the rubbing action accentuates the wear on crushing surfaces.

Under frictional rolling, milling results. The basic objective of a milling system is either to remove the husk and the bran layers, and produce an edible, white rice kernel that is sufficiently milled and free of impurities, or reduce the grain to powdery form. Milling operation is a crucial step in post-production process of grain crops such as rice. Depending on the requirements of the customer, the grain should have a minimum of broken kernels. Many mills combine different methods. Those commonly found are roller mills, claw mill and pellet mills.

Roller mill

The roller mill uses a pair of opposed toothed rollers that rotate simultaneously in opposite directions and at different speeds to grind the feed. If the rollers rotate at the same speed, compression is the primary force used, if the rollers rotate at different speeds, shearing and compressions are the primary forces used. If the rollers are grooved, a tearing or grinding components is introduced. Roller mills are mainly used for grinding oil cakes.

Advantages of roller mill

Advantages of roller mill include:

1. Energy efficient,
2. Uniform particle size distribution and
3. Little noise and dust generation.

Disadvantages of roller mill

Disadvantages include

1. Has little or no effect on fiber,
2. Particles tend to be irregular in shape and dimension,
3. May have high initial cost (depends on system design) and
4. When required, maintenance can be expensive.

Claw mills

Claw mills hit and grind material with claws fixed in a rotating disc, and are suitable for concentrate grinding because of compact structure, small volume and light weight. The structure of a claw mill with its feeding, grinding and discharging parts consist of a feed hopper, a feed control door and a feed tube. The grinding part consists of a rotary serrated disk, a stationary serrated disk and a ring screen. Claws are fixed on the rotary and the stationary disks alternatively. The discharging part is a tube situated in the bottom of the machine.

Pellet mill

The pellet mill consists of specially made rollers and dies which use special binders to produce high pressure to palletize the feed for poultry, fish etc. They are available in both vertical as well as horizontal designs and have high capacity. The rollers and dies are made of special alloy steel, which are hardened to the required properties for reducing wear and tear. It is useful for making pellets for the feeding needs of poultry, cattle, pigs or aqua feed.

Disc mill

The disc mill is another option for particle reduction. Grain is introduced to a chamber between one stationary disc and one revolving disc.A typical disc feeding component comprises of two grinding plates; one fixed on the machine while the other is attached to the driving shaft which is connected to driven pulley. The material come in-between the two plates as it is driven by an engine or induction motor. The grain is crushed as it moves from the center to the outside of the chamber through reducing clearance between the discs.

Adjusting the distance between the two discs controls the particle size, and this can be automated to allow different clearances for different feed grains in the ration.

Figure 5-37: Disc milling machine

This is an attractive capability, but adjustment between grains reportedly requires about 2.5 minutes, which significantly increases batching time. As a result, many disc mill owners set their disc clearance based on the ration and run all grain through that same setting, with the same result as running all grains through one size of screen on a hammer mill. The discs are vulnerable to damage if foreign items (stones, sand, and metal) enter the chamber, and due to the high cost of replacement discs, special cleaners are recommended for grain going into a disc mill.

Plate mill

A plate mill consists of a circular chamber made of cast iron or steel within which two plates with a narrow gap between them mounted face to face. The plates are grooved in order to provide a shear mechanism (Figure 5-38). When grains are introduced into the centre of the mill, the plates shear the grains between them. One of the plates rotates and the grains revolve, working their way to the outer edge of the plate before dropping by gravity into a holding sack below.

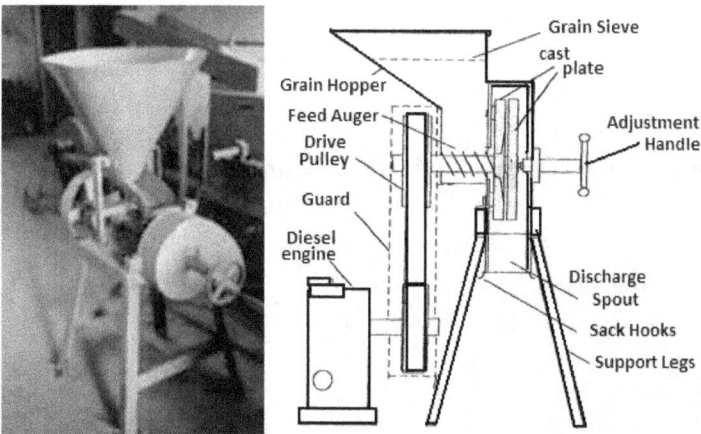

Figure 5-38: Locally made plate mill and typical construction features

The plates are usually about 200–300 mm in diameter, aligned in a vertical direction; however, horizontal alignment is more convenient when the mill is run by a diesel engine. Plate mills can run as fast as possible but normally at about 2 500–3 500 revolutions/minute, as overheating of the plates limits the speed of the mill. The speed of mill is not a critical factor to the mechanism of grinding. Plate mills operate more effectively with soft and moist grains that shear easily than with hard and brittle grains. It is common in West Africa to add water at the time of grinding. The milled product has to be used very quickly in order to prevent fermentation. Plate mills are popular in West Africa and the Sudan and operate with a greater shear than compression.

Effects of mill type on animal performance

Increased particle size uniformity and (or) using a roller mill to grind grain improved apparent nutrient digestibility in diets and decreased undesirable changes in stomach morphology (Wondral et al., 2011). Further research is warranted to verify these responses in swine herds having problems with high incidence of stomach lesions. It is difficult to attribute the benefits in apparent nutrient digestibility from fine grinding of cereal grains to decreased mean particle size alone, because increased uniformity of particle size is related to decrease mean particle size. Research results (Wondral et al., 2011) suggests that mill type had inconsistent effects on growth performance, but more uniform particle sizes consistently gave greater nutrient digestibility.

As for effects of roller mills and hammer mills on growth performance of e.g. pigs, some also suggest that the more uniform particle size (i.e., lower standard deviation of the mean particle size) achieved with roller-mill grinding has nutritional significance (Wondra et al., 1995b) studied the effects of particle size uniformity and milling with a hammer mill or roller mill on growth and other performance characteristics in finishing pigs.

Feed quality measurement

There are many quantitative methods for measuring feed quality as product quality concerns have always been a part of feed manufacturing. The physical traits (appearance, feel, handling characteristics) will always influence the feed buying customer. The grind produced by a roller mill is very uniform producing a finished product(s) that have an excellent physical appearance.

There are many benefits to producing a feed with minimal particle size variation. There are many quantitative methods for measuring feed quality as product quality concerns have always been a part of feed manufacturing.

5.6   Mixing materials of agricultural materials

Mixing involves the putting together of two or more substances so that the particles of each are diffused among those of the others. It is one of the processes involved in feed preparation and must be attended to with care because improper mixing of feed ingredients result in unbalanced rations that undernourished livestock.

The objective of the mixing process is to produce feed in which nutrients and medication are uniformly distributed. Well mixed feed enhances animal performance and is an essential step in complying with FDA good manufacturing process. The state in which any sample removed from the mixture has exactly the same composition is known as "perfect" mixing.

Feed mixers

There are certainly design differences in the mixers available. Design changes are driven by market and consumer demand. Take for example the consumer demand that a mixer should be able to handle the addition of long dry hay into the ration. This particular design change could cause another potential problem with misuse; particle size reduction with too long mixing time.

Mixer design is still primarily a trial and error process with due consideration given to prior experience. The manufacturer selects a specific mixer design that is expected to perform and field tests, determine design changes and their effect on the mix. In a summary paper on mixer design, these design and testing issues were identified:

1.   Mixer design (type, geometry, power, time, speed, efficiency)
2.   Define material changes (particle size reduction)
3.   Define standards for comparison of mixers
4.   Classify and measure degree of mixing (determine the quality of the mix)
5.   Describe the mixing process
6.   Correlate quality of mix with respect to time

All of these are good research projects, and some manufacturers and researchers may have answers to some of these problems. Other issues may not ever be solved. As an industry, there is no effort to coordinate the research efforts or develop standards for testing.

Types of feed mixers

Several mixers are in use today though some disadvantages make them inefficient. Mixers in the market fall into several general design categories and the followings describe the basic mixer designs (Kammel, 1998)

a. Horizontal auger mixers

This mixer uses one, two, three or four augers to churn the feed in a hopper (Figure 5-39). The feed moves along the flighting of the auger(s). In one and two auger mixers the flighting moves feed toward the middle of the mixer and it bubbles to the top toward the sides and back down to the augers. Feed is also moved to the discharge door from both ends of the mixer.

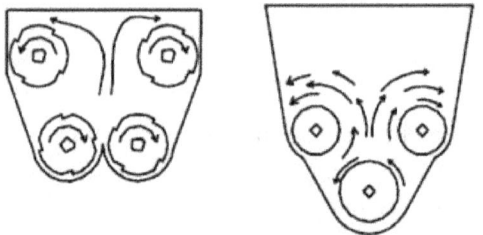

Horizontal four auger   Horizontal three auger

Figure 5-39: Horizontal mixers

In 3 and 4 auger mixers, one or two counter-rotating auger(s) and/or flighting moves feed in the opposite direction of the other augers. Feed moves from end to end and from bottom to top. The feed eventually moves toward the discharge door and is unloaded when the door is opened. In many mixer designs, notched auger flighting and/or knife sections attached to the auger flighting provide the ability to cut or tear long hay into 3-4" pieces and incorporate it into the ration. Design differences in these mixers include rotation speed of the augers, auger diameters, and auger flighting design.

b. Reel mixer

The mixer combines a set of augers and a reel similar to a combine reel in a hopper (Figure 5-40). Feed is lifted and tumbled by the reel moving it to the rotating augers, which provide a mixing action, moves feed from end to end, and to the discharge door. Knife sections on the auger flights cut or tear long dry hay into 3-4 inch pieces and incorporate it into the ration. An optional hay pan allows the hay to be metered into the mixer providing the ability to break up large portions of dry hay or baleage.

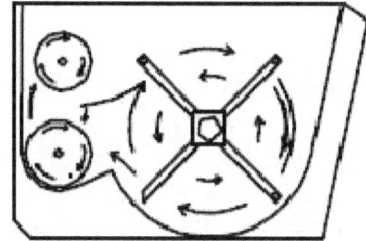

Figure 5-40: Reel mixer mechanism

c. *Tumble mixer*

The mixer is a large drum with spirals and/or pans on the interior circumference of the drum to lift and tumble the ration. (Figure 5-41) A central auger moves feed from end to end and to the discharge. A large part of the drum opens like a door to allow loading with a skid-steer or loader bucket.

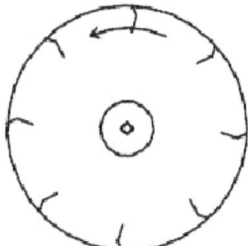

Figure 5-41: Tumble mixer mechanism

d. *Chain and paddle*

The mixer uses a tub or box containing a chain and paddles or slat conveyor to tumble the feed ingredients within the tub end to end. (Figure 5-42) An auger at the front of the mixer provides additional mixing and moves material to the discharge.

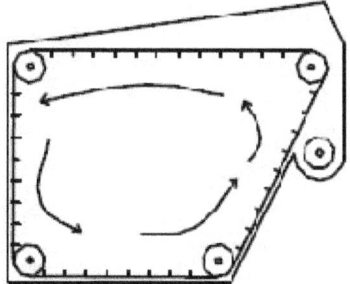

Figure 5-42: Chain and paddle mixer mechanism

e. Vertical screw mixer

The mixer consists of a large tub with a single vertical tapered screw centered in the tub. (Figure 5-43) A planetary gearbox and transmission drives the screw. Knife sections are attached to the flighting to cut material. Movable shear or restrictor plates on the tub wall provide a shear surface, increasing the ability to process and reduce the particle size of large packages of hay. These units can process rations with almost 100% dry hay. No prior processing of hay is required.

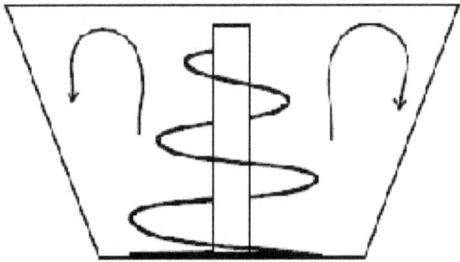

Figure 5-43: Vertical mixer mechanism

f. Mixer cart

Mixing carts are scaled down versions of some of the designs discussed above. There are chain and paddle, tumble, and reel mixer cart designs on the market. Sizes range from 40-80 C.F. They are usually powered by a small 8-18 HP gas engine. They are used where a smaller volume of feed is needed. The mixing cart can feed approximately 12-24 cows per batch mixed. It is a popular option for small herds. Research herds also find them useful for feeding cows on nutritional trials.

Factors affecting mixer performance

The following factors have been identified to have significant impact on the performance of feed miser operations.

a. Mixing time

The mixing time necessary to produce a homogenous distribution of feed ingredients should be measured. The type of mixer design (ribbon, paddle or auger) will have an impact on the mixing time needed to produce a homogenous mix. Each mixer should be "tuned" to its proper revolutions per minute (RPM) for optimum ingredient dispersion. Manufacturer's recommended mixing times range from 3-6 minutes depending on the mixer design etc.

b. Sequence of ingredient addition

The sequence of ingredient addition also determines the pattern of ingredient dispersion in the mixing process (Herrman and Behnke, 1994). Different types of ingredients may have a different flow pattern within a mixer at similar RPM's. In general the higher the speed (RPM), the more efficient the feed dispersion pattern in the mixer (Wilcox and Unruh, 1986). Many mixers may have dead spots, where small amounts of ingredients may not be readily incorporated into the feed.

The addition of liquid ingredients (fats, oils, molasses, liquid chlorine chloride, alimet and other liquids) into the mixer is a common practice in many milling operations. A spray bar installed at the top of the mixer is the best way to introduce liquid ingredients. Prior to introducing any liquids into the mixer, the dry ingredients should be adequately mixed. Premature liquid addition tends to impede the transport of micronutrients and may even agglomerate the fine particles into "snowballs".

c. Mixing capacity

The mixing capacity per batch of feed of a mixer is the total volume of the mixing compartment. Mixers are rated from 60-90% of the stock/capacity of the mixer depending on the manufacturer. Total mixed rations (TMR) have becomes the major feeding system of the diary industry. Emptying the mixer completely during delivering will help the processor to avoid cross contamination because if the mixed feed is cross contaminated, it will affect the animal performance which will be a problem during the production/output per year

d. Other factors

Although insufficient mixing time and filling of the mixer beyond the rated capacity are often implicated as common sources of variation in finished feed, other factors such as particle size and shape of the ingredients, ingredient density, static charge, sequence of ingredient addition, worn, altered, or broken equipment, improper mixer adjustment, poor mixer designed, and cleanliness can affect the mixer performance (Wilcox and Balding, 1986; Wicker and Poole, 1991, Albert, 2011).

Feed mixing operation

Mixing is an operation basic to feed manufacturing and one operation necessary to define a feed mill. It is recognized as an empirical unit operation, which means that it is more of an art than a science and must be learned by experience (Köster, 2003). The objective of feed mixing is to start with a certain assortment of ingredients called a formula, totalling some definite weight. These ingredients must be combined through a mixing process to be fed as a complete diet.

From a feed manufacturing perspective, the optimum mixing procedure would require minimal inputs of time, electricity, and labour. At the same time, if growth, production and health of animals need to be optimized they should receive a balanced diet that supplies nutrients and feed additives at the desired concentrations.

In batch mixing operation, the ration first added followed by feed ingredients one at a time until the required weight of each specific ingredient is reached and the batch is complete. The order of addition of feeds can affect the mixing ability and/or time of mixing (Wilcox and Unruhm 1986). In addition, the loading point of the mixer may affect the time required to get a complete mix. Follow manufacturer's recommendations on addition order of feeds, and the recommended mixing time.

Nutrient uniformity and assessment

Beumer (1991) cited uniformity as one of the most important quality aspects in feed production. Thus a standard is needed to indicate adequate, but minimal, mix uniformity. In reality, there is currently no official testing procedure to describe mix uniformity. However, common tests that have been developed to use either an indigenous nutrient (e.g., chloride or sodium ion) or an added marker (e.g. coloured iron filings) to evaluate uniformity (Eisenberg, 1992).

Once random samples, collected throughout the batch of feed, have been analyzed, a coefficient of variation (CV) is determined for the distribution of a specific nutrient or marker within the feed and to ascertain the accuracy of the mixer. One general assumption is that feed mixers of any possible size, shape, design and configuration must achieve a coefficient of variation (CV) of 10% or less as the most important indicator of mixing ability and proper functioning (McEllhineey et al., 1991).

A CV of 10% has become the accepted degree of variation separating uniform from non-uniform mixes (Duncan, 1973; Beumer, 1991; Wicker and Poole, 1991). A CV% over 10% is unacceptable and tests must be re-run. A why-why analysis must be performed if the result of the second test on percentage coefficient of variation (CV %) is above 10% (Albert, 2011).

To calculate CV%, the following equation is used:

$$CV\% = \frac{\bar{\sigma}}{\bar{x}} \text{... ... ... .5.2}$$

Where $\bar{\sigma}$ = Standard deviation, $\bar{x}$ = Mean

Further reading

Albert A., 2011. Feed Manufacturing History Association of Kenya Feed Manufacturers. Updated on Friday, 13 May 2011http://www.akefema.com/index.php Copyright © 2010 Association of Kenya Feed Manufacturers

Beumer, I.H. 1991. Quality assurance as a tool to reduce losses in animal feed production. Adv. Feed Tech. 6:6-23.

Bello R. S., 2012. *Agricultural Machinery & Mechanization*. 7290 B. Investment Drive Charl 7290 B. Investment Drive Charl createspace ISBN-13: 978-1456328764. https://www.createspace.com/3497673 (344 pages)

Clarke B. and Rottger A., 2006. Small mills in Africa Selection, installation and operation of equipment Agricultural and Food Engineering Working Document 5 pp 5-8

Duncan, M.S. 1973. Nutrient variation: affect on quality control and animal performance. Ph.D. dissertation, Kansas State University, Manhattan.

Ensminger, M. E., J. E. Oldfield, and W. W. Heinemann. 1990. Feeds and Nutrition. 2nd ed. Ensminger Publishing Co., Clovis, CA.

Hermann T. and Behnke K. (1994). Testing Mixer Performance. Kansas State University Agricultural Experiment Station and Cooperative Extension Service pp 1-4.

Ige M.T and Finner M. F. (1976): Optimization of performance of the cylinder type forage harvester cutter head. Transactions of ASAE 38 (6): 1655-1658

Kammel D. W., 1998. Design selection and use of TMR mixers.doc 11/25/98

Köster Hinner, 2003. Improved animal performance through feed processing technology. Paper presented at AFMA's annual symposium on 22 August 2003. Pp 20-55

Mani, Lope S. Tabil G. Sokhansanji S., (2004). Grinding performance and physical properties of wheat and barley straws corn Stover and switch grass. Biomass and Bio-energy 27: 339-352.

McEllhiney, R. R. and C. Olentine. 1982. Problems with mixing. Feed International 3(5):34-38.

Munoz-Hernandez G., Gonzalez-Valadez M., and Dominguez-Dominguez J. (2004). "An Easy Way to Determine the Working Parameters of the Mechanical Densification Process".

Nelson, S. O. (1980). Moisture-dependent kernel and bulk density for wheat and corn. Trans. ASAE 23:139-143.

Nikolov, S. (2004): Modeling and simulation of particle breakage in impact crushers. International Journal of Mineral Processing, 74 (5): 219-225

Ogunlowo A. S. and Bello R. S., 2005. Design, construction and performance evaluation of cowpea Thresher. *Journal of Agric. Engineering and Technology (JAET)* Vol. 13 Pp 83-89

Reznicek R., 1988. Physical properties of Agricultural materials and their influence on Design and performance of Agricultural machines and technologies. Opening speech of the third international PPAM conference, Prague, Czechoslovakia. Hampshire Publisher USA.

Smith, A.E. and Wilkes, M.S., 1994. farm Machinery and Equipment, 4th edition, Mc Graw-Hill Publishing Company Ltd. Pp 374 – 375.

Whitney R.W, Porterfield J.G., 1968. Particle separation in a pneumatic covering system. Trans. of the ASAE Vol 11 No4 477-479

Wicker, D. L., and D. R. Poole. 1991. How is your mixer performing? Feed Management 42(9):40-44.

Wilcox, R.A., D.L. Unruh. (1986). Feed Mixing Times and Feed Mixers. Cooperative Extension Services. Kansas State University.

CHAPTER 6

Agricultural & Bioprocess Technologies

6. Introduction

Food, fiber, and timber are only the beginning of a long list of products that benefit from efficient use of our natural resources. Conversion of agricultural raw materials or products into finished consumer goods involve, harvesting, transporting, handling, storage, processing and packaging. Food, fiber, and timber are only the beginning of a long list of such products that benefit from efficient use of our natural resources (land, water etc). The list includes biomass fuels, biodegradable packaging materials, pharmaceutical and other products.

Figure 6-1: Food processing plant

Agricultural technological processes are various operations that agricultural materials are subjected to in attempt to improve its quality for direct consumption or for further processes. A wide range of machinery and equipment are required, some of which are reported in the following sections.

6.1 Grain processing

Rice milling

Rice milling is divided into two parts namely; husking process in which husks are removed from paddy, and the polishing (whitening) process in which white rice is obtained by scraping the top layer (bran layer) off the husked brown rice obtained by hulling. Rice milling operation can be carried out in the following stages

Single stage mill: The two operations mentioned above are accomplished by one single machine in one process. The rice grains are rubbed together and this friction removes the husks along with the bran layer of brown rice. Engel berg type hullers belong to this type of machine. These machines are no longer produced in rice nations as Japan today.

Double-stage mills: The process in this type of machine involves tow clearly separate operational units; husking and polishing. The husking unit employs a rubber roll system, and the polishing unit is mostly of the friction type. The presence of High speed turbo fan assists in cooling and bran removal.

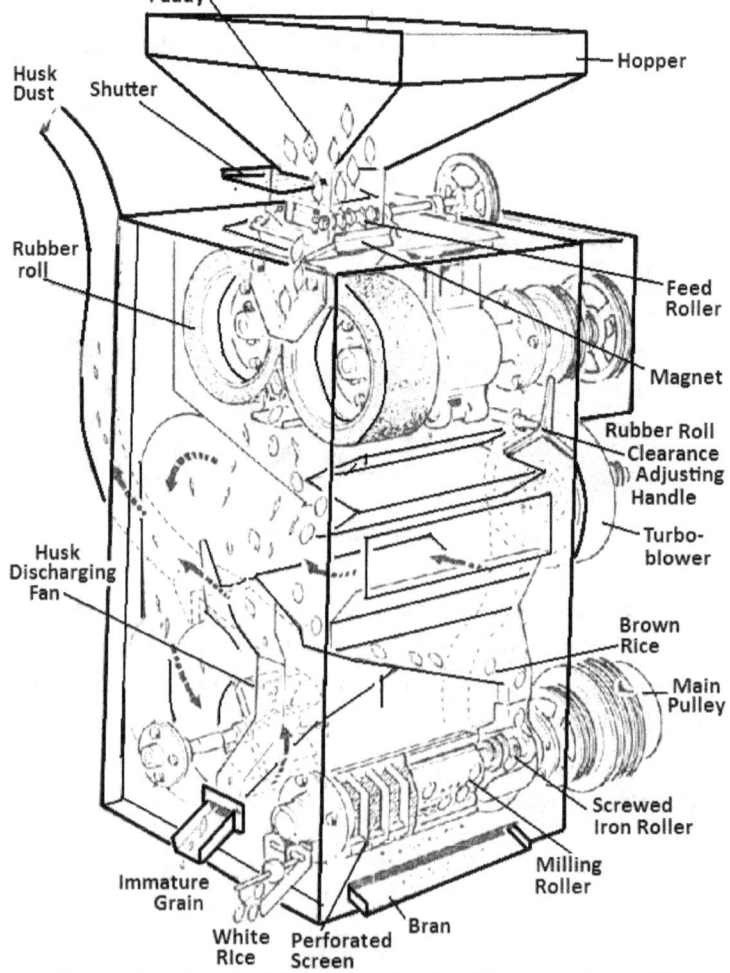

Figure 6-2: Internal features of a double stage rice mill.

Multi-stage mills: These are larger capacity mills, which are capable of treating a ton or more of paddy per hour, employ a further sub-divisions of processes usually; cleaning, husking paddy separation, whitening and post-treatment grading, cleaning, packaging weighing etc.

Seed grain polisher

This machine consists of stainless steel hexagonal drum supported on heavy-duty rollers. It has variable angle adjustment device for tilting of the drum, discharge outlet, frame and electrical system. All the assemblies are mounted on steel frame. The operation of the polisher is similar to the concrete mixer. Rotational speed of the drum can be adjusted. The grains are placed in the hexagonal drum and on rotation tumble the grain, which get polished due to rubbing action. Aspirator fan can be added for removing dust, husk, etc.

Grain destoner

This is a high performance machine, which removes stones and other heavy trash from the sample by separation according to the difference in specific weight. Materials enter the machine through the inlet air lock (which prevents ingress of air with the product. It then moves from the preliminary sieve to the trapezoidal working table. The working table is covered with a woven wire mesh bed, through which air flows from below. Heavy impurities such as stones are not lifted by the air cushion and stay on the mesh bed. The vibratory motion of the table causes this heavy material to be transported to the beginning of the regulating plate at the far end of the machine (Figure 6-3).

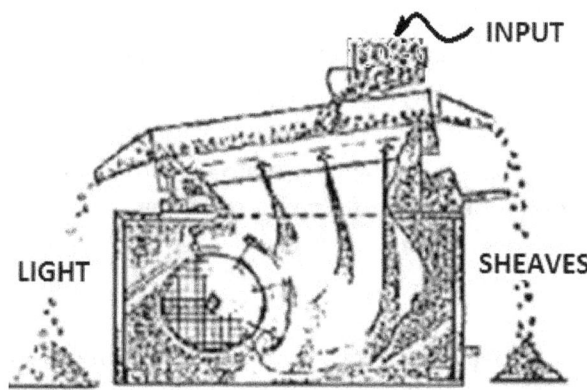

Figure 6-3: Grain destoner

Under this plate, the air flows in an opposite direction to the movement of grain on the working table. Due to this direction of air motion stones are separated from the product and moved to the stones outlet. The machine is fully adjustable to achieve the optimum separation in every case; the inclination of the working table, the stroke of the vibratory motor and the volume of the air used can all be regulated.

Advantages of destoner

Advantages of destining machine include

1. Smooth and silent operation,
2. Powered by electric vibratory motor for negligible maintenance,
3. Easy to adjust for optimum performance,
4. Has high capacity from compact body size and low power consumption.

Seed -grain dryer batch type

The dryer is indirect heating type, simple to install, operate and maintain. The dryer frame and body are made from mild steel sections and. It consists of portable heating unit with automatic jet oil burners with automatic ignition, centrifugal type dynamically balanced high capacity fan for generating hot air, chimney for discharging the effluent gases, sensitive thermostats for controlling the temperature, drying chamber with trays, discharge spouts and other controls such as cut-off, restart, solenoid valve, photo cell, sequence controller etc. The heating unit is mounted on wheels for easy portability.

Grain handling equipment

Grains include the cereal grains (maize, rice, wheat, sorghum, barley, oats, millet), the oil seeds (soybeans, sunflower seed, canola), and the pulses (edible beans). Although grain specie may require a particular postharvest operation, the fundamentals of the threshing, cleaning, drying, and storage of grains are sufficiently similar to warrant a general description of the various postharvest operations.

Threshing

Threshing involves the detachment of grain kernel from the panicles is one of the most critical post-harvest operations. Threshing of grain is achieved traditionally (manually) or mechanically. Manual threshing is mostly applied using cocoa bags, or spreading large clean cloth or tarpaulin on the floor, laying a bundle of grain head on the cloth and beating with heavy sticks and clubs.

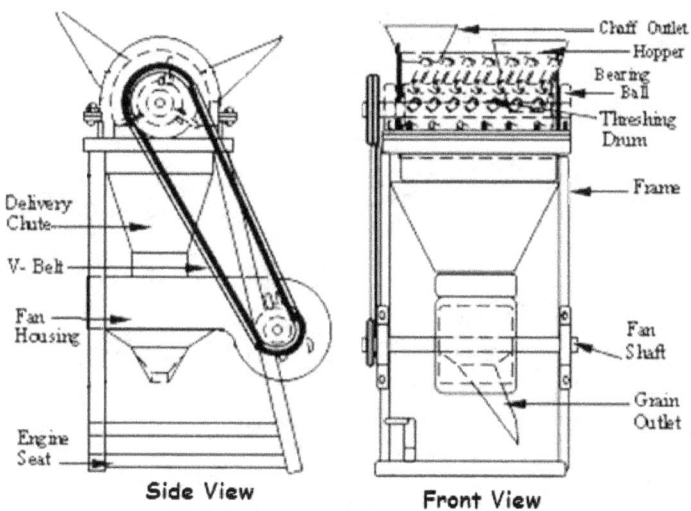

Figure 6-4: Components of a grain thresher *(Bello and Ogunlowo, 2005)*

Mechanical threshing of grain employs various thresher mechanisms such as spike tooth, rasp bar and angle-bar mechanisms to thresh grains (Claude Culpin 1987). Cowpea is a most susceptible leguminous crop to impact loading due to the di-cot nature of its kernels and is most affected in threshing with iron beaters.

Alternatively, animals (horses and bullocks) are allowed to trample on them (Igbeka and Oluleye, 1986). Grain losses are experienced during threshing. The conventional tangential threshing unit threshes mostly by impact; other threshing devices such as rotary threshing units act more by rubbing. A spike-tooth threshing cylinder offers some advantages for rice.

Grain cleaning

Once the seed is separated from the fruit it is ready for cleaning. There are a number of methods that include sieving, blowing, winnowing, flotation or imbibing the seed followed by gravity separation. Seed is rarely clean enough for immediate storage following collection. Most collections require harvesting of fruit that must then be processed by drying or depulping, extraction of the seed from the fruit, further cleaning and fumigation.

Winnowing

This procedure makes use of air current to separate seed from impurities through differences in weight, resistance to flow of air (volume or shape), and the velocity at which the air moves.

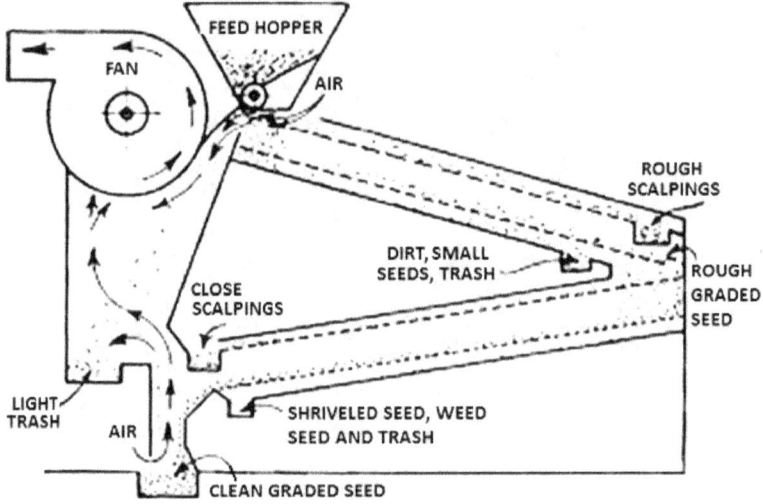

Figure 6-5: Air and screen seed-cleaner machine

Sieving

This method is most effective for the majority of seeds including pepper, tomato, eucalypts, etc. It is normally the only method available for cleaning in the field. Sieves come in a range of sizes, apertures with sieve material made from either perforated plate or woven wire. For small seeds, 20 cm diameter laboratory sieves with a wide range of aperture sizes are normally used whilst large sieves (50–80 cm in diameter) are preferred for large seedlots especially during the initial stages of cleaning.

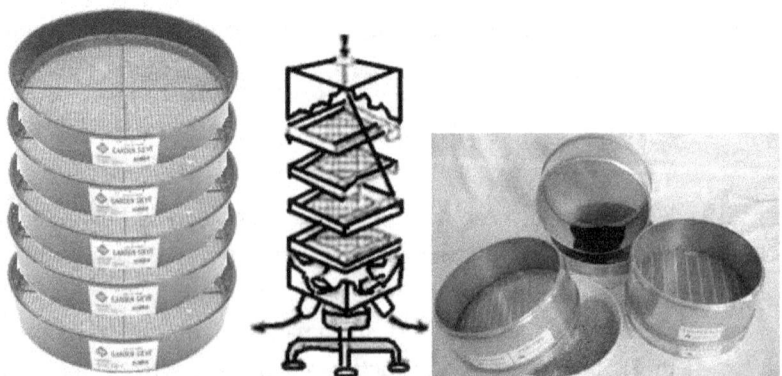

Figure 6-6: Sieves

Mesh sizes in common use vary from 500 micron to 4 mm to 12 mm. A combination of a large and small aperture sieve can be effective in removing both large and fine particles. An example of this is cleaning threshed grains where the chaff can be easily separated from the seed by use of a fine sieve while a larger aperture sieve removes the larger impurities.

Grain flour separator

The grain flour separator is a motorized unit for sieving different sizes of milled powder. It consists of four sieves of different mesh sizes, one over the other fig 7.41. It is mounted on a frame and gets its oscillatory motion form a cam mechanism. The prime mover uses is a 1hp single-phase electric motor. Milled wheat flour can be separated into bran. It is suitable for milled wheat, grain and soy flour, It consists of hopper separation chamber with appropriate screen, shaking unit and outlets.

Grain quality

Grain quality is interpreted differently by various end-users. To the livestock producer, the nutritive value of grain is important. To the cereal/food manufacturer, some grain physical property such as the breakage susceptibility may be of significance. And to the seed producer, only the seed viability is of interest. Regardless of the particular grain-quality criterion, the post harvest operations to which a grain sample is subjected determine its value.

Grain quality factors

The property of a lot of grain that determines its market value includes:

1. *Physical property:* Physical properties such as the kernel damage, the bulk density, or a nutritive value such as the crude protein content could determine its marketing value.
2. *Foreign or fine material:* The foreign and fine material in a grain sample is defined as the particles passing through a screen of specified design, plus the large pieces of extraneous matter. For maize, a 4.76-mm round-hole sieve is used, for wheat a 1.98-mm round-hole sieve, and for soybeans a 3.18-mm round-hole sieve. The foreign and fine material in a grain sample usually is measured in a dockage tester, which basically is a mechanical sieve shaker. The foreign and fine material content is expressed in terms of the percentage in weight of the original sample.
3. *Kernel damage:* The kernel-damage level of a grain sample is expressed as the percentage in weight of the original sample. Damaged grain kernels include broken kernels, heat-damaged kernels, discolored kernels, or shrunken kernels. In grading a grain sample, the grader distinguishes between heat damage from high-temperature drying and heat damage resulting from mold activity. This latter type of kernel damage is counted under the category of total damage and not of heat damage. Under the U.S. grain standards; the total damage category includes

also the sprouted, germ-damaged, weather-damaged, molded, broken, and insect-damaged kernels.

4. *Stress cracks:* Stress cracks are small fissures in the internal endosperm of maize and rice kernels and are caused by large moisture-content gradients (and the resulting compressive and tensile stresses) within the kernels due to rapid moisture absorptions or adsorption. Stress cracking often does not occur until 24 hours after drying or moisture adsorption. Excessive stress cracking results in a high percentage of maize kernels breaking during handling operations, and in a low value of the head yield of rice after milling. The percentage of stress-cracked kernels in lots of maize and rice is determined by candling of 100 kernels over a 100- to 150-W light source.

5. *Breakage susceptibility:* The breakage susceptibility of a grain sample is an indicator of the likelihood of the kernels to break up during handling and transport. It can be considered as an indirect measure of the number and size of the stress cracks in the kernels of a grain sample. The concept of breakage susceptibility is used for evaluating the quality of maize, mainly by researchers in comparing different maize hybrids and various maize-drying systems

6. *Nutritive attributes:* The nutritive value of grains is of importance to humans and animals alike. The significant nutritive properties of grains include: Ash, Crude fiber, Crude protein, Ether extract, Gross energy, Metabolizable energy, Nitrogen-free extract, and total digestible nutrients Ensminger (1991)

6.2   Transport of agricultural materials

Conveyor is one of the grain-handling equipments available for any situation under which grain has to be transported from one location to another. The grain- flow rate, distance, angle of inclination, available space, environment, and economics influences conveyor design and operating parameters.

The physical characteristics of the material to be handled must be known before the appropriate conveying system can be selected. In particular, the following properties are relevant for agricultural products: moisture content, average weight per unit volume, angle of repose, and particle size

Types of conveyors

Four types most commonly used for commercial and farm applications are belt, screw, bucket, and pneumatic conveyors.

Belt conveyor

The simplest belt conveyor consists of an endless flat or troughed belt wrapped around two rotating drums. Bulk materials are transported horizontally on the surface of the belt or can be inclined or declined within product-specific limits.

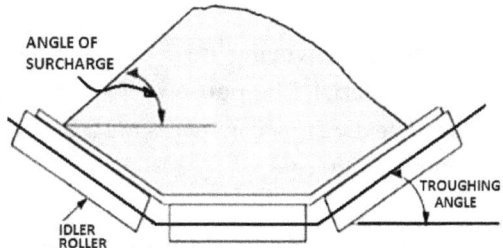

Figure 6-7: Cross-section of troughed-belt conveyor

Support for the belt is provided by a head (drive) and a tail pulley, and a series of idler rollers. The belt conveyor structure supports the entire system and maintains belt alignment. A typical belt-conveyor design is shown in Figure 6-6 below

Screw conveyors

Screw conveyors have proven to be one of the most reliable and cost-effective ways for conveying bulk materials. Since Archimedes invented the screw conveyor back in 267 B.C., it has been a versatile machine that can handle a wide variety of materials, from dry and free flowing, to wet and sluggish. A screw conveyor, or auger, consists of a circular or U-shaped tube in which a helix rotates, as shown in Figure below. Grain is pushed along the bottom of the tube by the helix; thus the tube does not fill completely. Important parameters of the auger-the diameter, pitch, and exposed intake length-are indicated on the figure.

Figure 6-8: Screw conveyors

Screw conveyors are used to convey any type of bulk material and are found in thousands of applications. The major industries that utilize screw conveyors are:

- Agriculture production chemicals
- Food processing

- Lumber and wood products among others

How does a screw conveyor work?

Screw conveyors are volumetric conveying devices. Each revolution of the screw discharges a fixed volume of material. The purpose of a screw conveyor is to transfer product from one point to the next. Screw conveyors are always control fed at the inlet by another conveyor or metering device.

The flow rate or capacity of a screw conveyor is measured in cubic feet per hour. If the capacity is given in lbs. per hour, tons per hour or bushels per hour, it is converted to cubic feet per hour. Since screw conveyors are control fed at the inlet, the cross-sectional trough loading is less than 100 percent. CEMA has developed standards for trough loading based on the material classification codes. Figure 6-9 shows the parameters necessary for sizing a screw conveyor.

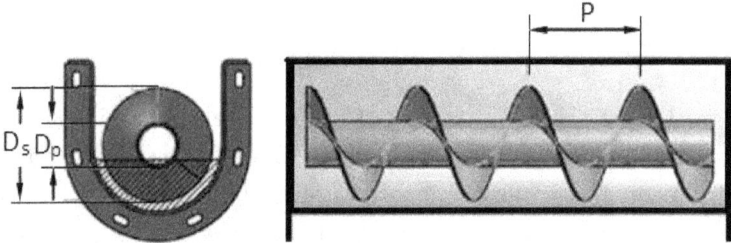

Figure 6-9: Screw conveyor configurations

The capacity calculation takes into account the outside diameter of the screw, the outside diameter of the pipe, the pitch of the screw and the trough loading. The calculation determines the capacity in cubic feet per hour that will be conveyed with each revolution per minute of screw rotation according to equation 6.1 (Bill Mecke, 2011)

$$\frac{C}{rpm} = \frac{0.7854(D_s^2 - D_p^2)PK60}{1728} = 0.0273(D_s^2 - D_p^2)PK \ldots\ldots\ldots.6.1$$

Where
   C = Capacity (cubic feet per hour)
   D_s = Diameter of screw (niches)
   D_p = Diameter of pipe (inches)
   K = Trough loading (%)
   P = Pitch of screw (inches)
   rpm = Speed (revolutions per minute)

Classifying bulk materials

Bulk materials are defined as goods that are handled in large quantities without the benefit of individual packaging. Bulk materials are conveyed, stored and processed to create the things we need to live, such as food products, plastics, building products, paper and thousands of finished goods.

The Conveyor Equipment Manufacturers Association (CEMA) classifies bulk materials by the following properties:

- Particle size
- Flowability
- Density
- Abrasiveness
- Other characteristics (corrosive, flammable, sticky)

The angle of repose of a material is the angle which the surface of a normal, freely formed pile makes to the horizontal.

Table 6-1: Characteristics of bulk materials

Flow	Very free	Free flowing	Average	Average	Sluggish
Angle of Surcharge	5°	10°	20°	25°	30°
Example					
Angle of Repose	0°-19°	20°-29°	30°-34°	35°-39°	40° up
Characteristics	Uniform size, small rounded particles, either very wet or very dry, such as dry silica sand, cement, wet concrete, etc.	Rounded, dry polished particles, of medium weight, such as whole grain and beans.	Irregular, granular or lumpy materials of medium weight, such as anthracite coal, cottonseed, meal, clay, etc.	Typical common materials such as bituminous coal, stone, most ores, etc.	Irregular, stringy, fibrous, interlocking material, such as wood chips, bagasse, tempered foundry sand, etc.

Source: http://www.cambelt.com/bulksolids.c

The angle of surcharge of a material is the angle to the horizontal which the surface of the material assumes while the material is at rest on a moving conveyor belt. This angle usually is 5° to 15° less than the angle of repose; though in some materials it may be as much as 20° less.

Bucket elevator

A bucket elevator is essentially a modified belt conveyor. Buckets are attached to an endless belt and convey the product vertically The components include the head section at the grain outlet (top), the boot section at the grain inlet (bottom), the supporting frame, a device to maintain correct belt tension, and the grain distributor. The enclosed section of the belt between the head and boot is called the leg; the up-leg and down-leg sections often are physically separated within individual enclosures.

Bucket elevators / conveyors are used for vertical transportation of powder, loose and granular materials of small sizes – materials to be consulted with the supplier – by means of buckets fixed on the transport belt in the upward direction in agricultural, food, chemical and wood-processing plants. Firmly adhesive and sticky materials cannot be transported.

Pneumatic conveyor

A pneumatic conveying system introduces grain into a moving air stream, which carries the grain to a single location or multiple locations. The main components include the blower, the transport tube, and the device to introduce the grain into the tubing at the grain source and out of the tubing at the grain destination. There are two types of pneumatic conveyors; positive-pressure and negative-pressure type as shown in (Figure 6-10).

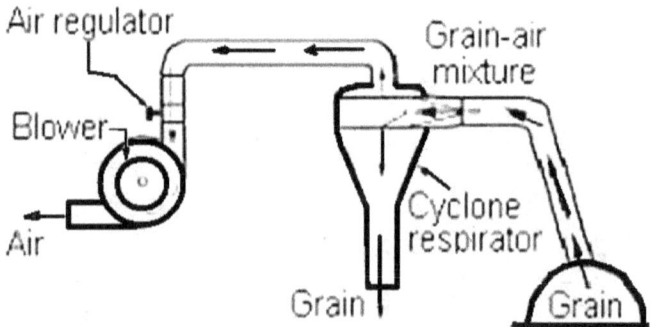

Figure 6-10: Negative-type pneumatic conveyors (CIGR, 1999)

6.3   Tuber crop process equipment

Potato destoner with prewasher

Before peeling, removing stones and careful washing of the potatoes is essential. In the cyclone destoner the rotating water convey the potatoes to the prewasher. The potatoes are washed in the drum. The stones that have a higher density will sink to the bottom, where they are collected or removed.

Figure 6-11: Prewasher with cyclone destoner for potato *(Rosenqvist Foodtech)*

The potatoes are then washed in a drum. Well washed potatoes minimizes the risk for sand, clay and other foreign particles, to stick or be trapped in gaskets, sealing etc., thus minimizes the need for maintenance.

The cyclone destoner has the optimal angle and water velocity to achieve effective separation of the stones and the potatoes. Stones are collected in a box or (optional) the destoner is supplied with a stone elevator for continuous removal of washed potato to a container. The washing drum is driven by an X-cross mounted belt. No bearing is in contact with water. The fresh water is added through a number of spray nozzles, placed along the drum. The water level of the washer is easy adjustable via an overflow valve arrangement.

The stone elevator (optional) can be mounted in any desired angle in elevation to the cyclone cone. It is built for capacities between 1000-20 000 kg/hr.

Prewasher advantage

1.   Effective stone separation
2.   Effective washing minimizes water consumption
3.   Simple maintenance due to reliable design

4. Hygienic design makes it easy to clean
5. Flexible stone elevator

Potato peeler

The potato-peeling machine consists of an indented cylinder, which has protruding rasps on the inner surface. It consists of mainframe, handle, rotating drum with notches, water inlet, top cover etc. Potatoes to be peeled are put inside the cylinder and rotated gradually. Due to the presence of the rasps, the peel gets removed. It is a batch type machine and about 8 kg of potatoes can be peeled in about 8-10 minutes. Water is used to wash away the tiny peels before the peeled potatoes are taken out of the machine. Since potatoes are free to rotate inside the cylinder, potatoes of assorted sizes can be fed into the machine.

Potato slicer

The potato slicer consists of a rotating disc, which carries radial knives and revolves on a vertical shaft. Over this disc, vertical cylinders are mounted for feeding the potatoes. It is mounted on a frame made of mild steel angle sections and rotated with a handle provided on the top. It consists of mainframe, four cylinder and pressing device. It can accommodate potato up to 75 mm diameter.

6.4 Fruit processing equipment

Fruit processing machines include complete plants required in various fruit handling operations. Examples include machines that process all types of fruits into juice, concentrate, jam and freshly prepared pieces, and packaging to suit.

Manual fruit/grain sorting

Incoming produce is placed in the sorting bin, sorted by one worker into the packing bin, and finally packed. The surface of the portable sorting table in Figure 5-11 below is constructed from canvas and has a radius of about 1 meter (about 3 feet). The edges are lined with a thin layer of foam to protect produce from bruising during sorting, and the slope from the center toward the sorter is set at 10 degrees. Produce can be dumped onto the table from a harvesting container, then sorted by size, colour and/or grade, and packed directly into shipping containers. Up to 4 sorters/packers can work comfortably side by side.

Figure 5-12: Sorting table (PHTRC, 1984)

Fruit grader

The fruit grader employs stepwise expanding pitch fruit grading mechanism based on the principle of changing the flap spacing along the length of movement of fruits. The grading mechanism has provision to separate fruits into four grades. The grading mechanism consists of two tracks of conveyor chains, matching sprockets, stainless steel flaps, and conveyor supports, flap space adjusting mechanism, sidewalls and fruit collecting chutes.

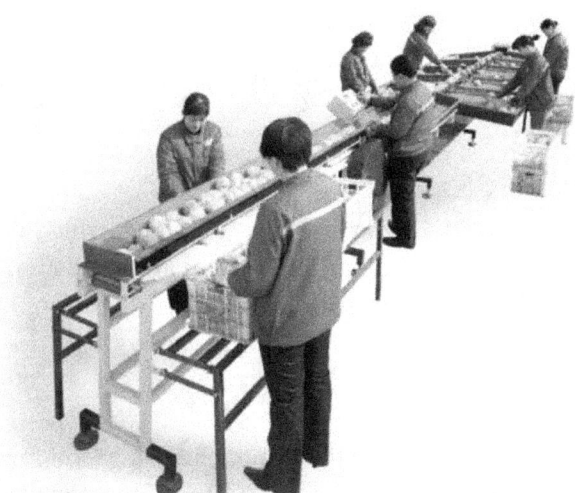

Figure 6-13: Fruit grading

The grader can be used to grapefruits such as apple, sweet lemon, oranges and larger size fruits. It is provided with elevator feeder for constant feeding of fruits in grading mechanism (Figure 6-13).

Figure 6-14: Weight sorter

Electronic colour sorter

The electronic colour sorter consists of a microprocessor-based roller feeding system provided with a three way scanning system which enables all round viewing of various grains, resulting in better quality of accepted and rejected product.

Figure 6-15: Fruit colour sorter

The machine is equipped with hybrid signal conditioning, which allows ultra fast and reliable signal processing. It is independent of variations in light intensity and background because of the auto-annulling feature.

Mechanical coconut dehusking

The traditional method of dehusking by farmers by using simple tools such as knife blade or spear is hampered by poor productivity. The need for mechanization is important for preparation of coconut in large-scale industrial processing due to labour shortages. In recent years, several machines were developed for dehusking of coconuts.

Figure 6-16: Mechanical coconut dehusking

This machine uses power from a 5-Hp engine and transmits to it to a hydraulic pump. The Hydraulic pump transmitted the power to the double spiked rollers which have 33 spikes on each roller. Both rollers are set in parallel position with 45 degree from base frame. The two rollers will rotate in opposite direction and scrapped the coconut skin and split the husk and nuts. The position of the frame ensures easy dropping of the coconut by gravity at one side and husk is spitted at the other side automatically.

Manually operated coconut splitting machine

Coconut is the world's largest nut with varying size and shell thickness. A nutcracker (Cocosplit) is mechanically used to split coconut with a firm blow from a large 1350 gram (3 pound) hammer. The juice can be saved by pre-draining, or by standing the Cocosplit in a small bucket or plastic dish.

Figure 6-17: Cocosplit mechanism (Mike Foale, 2012)

A stout elastic band wrapped around the nut to avoid loss of juice and restrains the half nuts from "exploding" out of the dish (Figure 6-17). The Cocosplit mechanism consists of a base plate for saving the juice, a circular cap, chisel point and spring. Locate the nut carefully on the cushion of the base-plate.

(Direct contact between shell and the metal base-plate may cause a jagged break in the shell). Ensure that the cutting tool blade rests on the centre of the top of nut, ensuring a split into equal halves. Both ends of the tool blade must be in contact with the shell.

Coconut punch & splitter

The manually operated splitting device is used for splitting dehusked coconuts. It consists of a pedestal, pivoted long handle, cutting knife and a platform. Lifting the handle raises the knife (Figure 6-17). The dehusked nut is placed on the platform and then the knife is lowered to cut the nut. The coconut water is collected is a pot or container placed at the base of platform after splitting the nut.

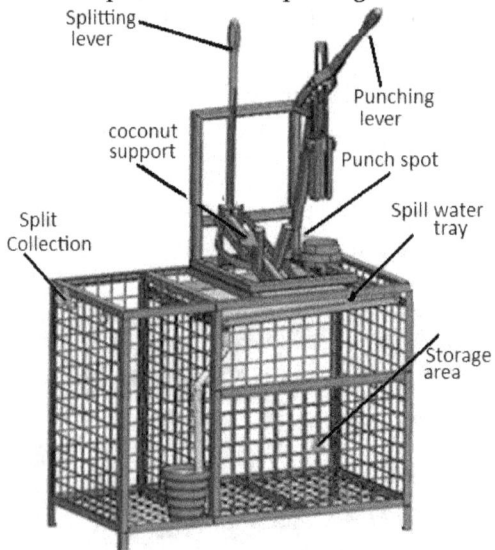

Figure 6-18: Manual punch and splitting machine, (IndiaMART, 2008)

Vacuum/specific gravity separator

Specific gravity separator is used for separating products that are of the same size but with a difference in specific weight. It can be used effectively to remove partially eaten, immature or broken seeds to ensure maximum quality of the final product. It may be used to separate and standardize coffee, peanuts, corn, peas, rice, wheat, sesame and other bold grains.

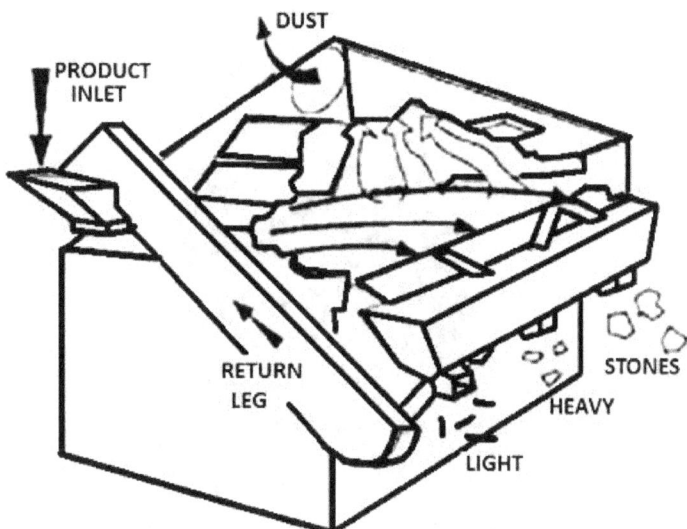

Figure 6-19: Vacuum gravity separator

The specific gravity separators have a long rectangular deck so that the products travel a longer distance resulting in cleaner separation of light and heavy particles and the lowest percentage of middling. Component parts of the gravity separator are an air blast fan, an air equalizing chamber, a perforated deck, variable-speed eccentric, deck rocker arms, a feeding or metering device, a check end-raise adjustment, and a deck side-tilt adjustment.

The end-raise adjustment varies the inclination of the discharge edge. The side-tilt adjusts the deck so that the back is higher than the front. One end of the deck is the discharge edge and the sides are the banking rails. The product flows over the vibrating deck in which pressurized air is forced through causing the material to stratify according to its specific weight. The heavier particles travel to the higher level and the lighter particles travel to the lower level of the deck.

Spiral gravity separator

The spiral separator classifies seed according to shape and ability to roll or slide. A simple machine has no moving parts and requires very little floor space (Figure 5-30). After it is placed under a feed hopper and the feed rate regulated, the spiral will operate steadily with little or no care, and with no power cost. It is light and relatively easy to move into or out of the processing line as needed. It may easily be taken to the seed location rather than moving the seed lot to it.

Figure 6-20: Spiral gravity separator

Some separations can be made with this unit that is not possible with air screen machines, length separators, and other seed cleaners. Basically, this separator consists of one or more steel flights wound on a central tube in the form of a spiral. The unit resembles an open screw conveyor standing in a vertical position. In most cases, spirals are built by local craftsmen. The critical measurements are the slope and inclination of the spiral "runs".

Soybean flaking machine

The flaking machine consists of three mild steel rollers (knurled and chromium plated surface), mainframe, hopper, stand, and collecting tray and drive mechanism. It is suitable for producing flakes from soybean, sorghum, maize and Bengal gram. It is driven by one hp single-phase electric motor through belt and pulleys and the drive to the rollers is by spur gears.

Blanching

Blanching is a process of de-activating enzymes through special heat treatment. Though it destroys enzyme, it retains the nutritional value and the taste of the vegetables. Blanching is applicable to different types of agricultural products like asparagus, soybean, potatoes, mushrooms, etc depending on the specificity of vegetables.

The blanching machine consists of a roller chain and an axial rod which goes through a specially made basket to form a conveyor set which convey the materials vertically, horizontally, or at an inclination. The vegetables or other ingredients are put inside the basket and then transported into the bucket with hot water for blanching. The hot flue gases pass to heat the water, which is stored in a concentric outer cylinder. The materials are placed in stainless steel perforated cages and immersed in the hot water.

Soybean blanching unit

Blanching of soybean is carried out to remove anti-nutritional ingredients. It is essentially a water heater with the provision for inserting and removing the containers of soybean. It has a central cylinder through which the hot flue gases pass to heat the water, which is stored in a concentric outer cylinder (Figure 5-20). The outer cylinder is insulated with asbestos rope to conserve heat. Soybean is placed in stainless steel perforated cages and immersed in the hot water. It consists of central heating cylinder, outer cylinder, asbestos rope insulation, stainless steel.

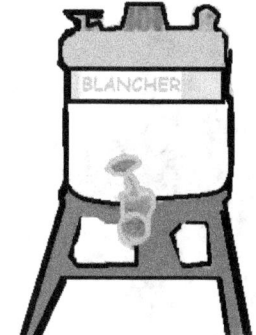

Figure 6-21: Soybean blanching unit

Working principle of peanut peeler

Peanut peeling machine adopts the wet peeling method to peel off the outer red skin from the peanut kernels. Roasted peanut kernels with outer red skin passes through the conveyor system into the feeder, and is distributed through tubes.

Figure 6-22: Peanut peeping machine (Agico group ltd.)

Then the peanuts automatically fall down the blanching chamber which is composed of two sand rollers. Under the actions of friction and rotation, the red skin is peeled away and then separated from the kernel by a suction fan connected by a cyclone system and finally blanched white peanut kernel fall down to the selecting conveyor.

Vegetable dryer

The vegetable dryer is specially designed for drying high moisture crops such as cauliflower, cabbage and onion etc to a low level of moisture content. Typical dryer designed by Solarflex shown in Figure 6-23 consists of a drying chamber, plenum chamber, heating chamber a blower. The drying trays are made of aluminum and have nylon mesh is provided for keeping the produce. Temperature control is achieved with the help of the thermostat provided in the unit. Moisture content of the produce can be reduced from 90% to 6% in a batch of 50 kg in 11-14 hours.

Figure 6-23: Solarflex horizontal vegetable dryer (drying rack)

Melon depodding machine

The mechanics of melon depodding include compression, shearing and impact. The main function of the melon-depodding machine is to break the melon pod and separate the seeds from the pod and the pulp.

The hopper serves as the feeding mechanism through which the melon pod is fed into the machine. The spike, through their impact force breaks the pod and the melon seeds, are separated from the pod and are subsequently collected through the outlets. The screw conveyor containing spikes conveys both the depodded melon pulp and seeds into the discharge outlets (Figure 6-24).

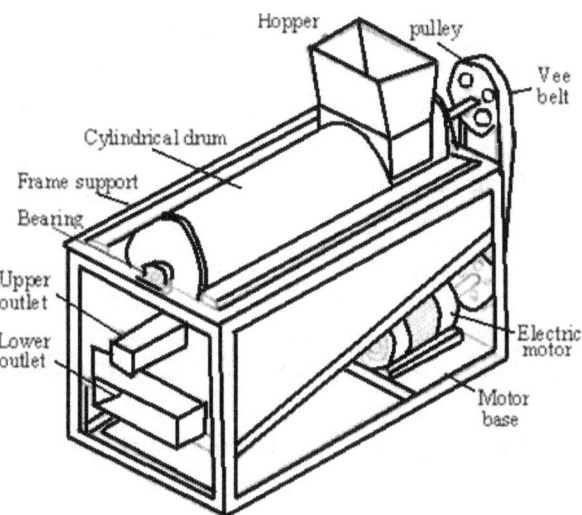

Figure 6-24: Melon depodding machine (Oloko and Agbetoye, 2006).

6.5 Pressing/extraction process

Plant products such as oil and tuber liquid can be extracted mechanically with an oil press, an expeller, or even with a wooden mortar and pestle—a traditional method that originated in India. Presses range from small, hand-driven models that an individual can build to power-driven commercial presses.

Figure 6-25: Oil extractors

Expellers have a rotating screw inside a horizontal cylinder that is capped at one end. The screw forces the seeds or nuts through the cylinder, gradually increasing the pressure. The material is heated by friction and/or electric heaters. The oil escapes from the cylinder through small holes or slots, and the pressed cake emerges from the end of the cylinder, once the cap is removed.

Both the pressure and temperature can be adjusted for different kinds of feedstock. The ram press uses a piston inside a cage to crush the seed and force out the oil. Oils can also be extracted with solvents, but solvent extraction is a complex operation. Oil is extracted from a number of fruits, nuts and seeds for use in cooking and soap making or as an ingredient in other foods such as baked or fried goods. Oil is a valuable product with universal demand, and the possible income from oil extraction is therefore often enough to justify the relatively high cost of setting up and running a small-scale oil milling business.

Oil clarification

Clarification removes contaminants, such as fine pulp, water, and resins from extracted oil. Oil clarification can be done in two ways as follows:

1. You can clarify oil by allowing it to stand undisturbed for a few days and then removing the upper layer. If it needs further clarification, filter the oil through a fine filter cloth.
2. You can heat the oil in a clarifier to drive off traces of water and destroy any bacteria. This consists of an oil drum placed above a fire (Figure 5-26).

Figure 6-26: Drum clarifier

The oil is heated briefly to 100°C to boil off any remaining traces of moisture. This is usually sufficient to meet the quality needs of customers and give a shelf life of several months when correctly packaged. However, the oil requires additional refining stages of de-gumming, neutralizing and de-colouring to have a similar quality to commercially refined oils, and these stages are difficult to complete at a small scale.

6.6 Products packaging and materials

Packaging as an agricultural operation can be an aid as well as a hindrance to obtaining maximum storage life and quality. Produce packaging ease material handling and provide protection to the produce when adequately provided for. Broadly speaking solid objects can be placed in wooden or wicker boxes or sacks and wooden barrels while liquids are transported in sealed containers, tins, barrels and drums for convenience.

Produce can be hand-packed to create an attractive pack, often using a fixed count of uniformly sized units. Packaging materials such as trays, cups, wraps, liners and pads may be added to help immobilize the produce. Simple mechanical packing systems often use the volume-fill method or tight-fill method, in which sorted produce is delivered into boxes, and then vibration settled. Most volume-fillers are designed to use weight as an estimate of volume, and final adjustments are done by hand (Kader, 2002).

Types of products packaging

There are two categories of packaging materials;

a. *Packaging materials requiring no protection*: Materials which do not need protection from rough handling can be shipped in bulk. Bags and baskets provide no protection to the product when stacked.
b. *Packaging materials requiring protection*: Goods requiring protection in transit need solid boxes round them. Boxes however come in a wide range of types; Cases are wooden boxes with solid sides, chests are light weight boxes often with the reinforcing framework on the inside, crates are open wooden framework boxes and cartons are cardboard boxes. Light but bulky material such as wool, jute and cotton is formed into bales and

Improved shipping methods such as containers, pallets and intermediate bulk containers have dominated the packaging scene. Packaging structures such as waxed cartons, wooden crates or rigid plastic containers etc. are preferable to bags or open baskets, since sometimes locally constructed containers can be strengthened or lined to provide added protection to produce.

Description of packing containers

Sacks: Sacks are often used to package produce, since they tend to be inexpensive and readily available. Grains and powdery materials are packed in sacks that are sealed and are used as shipping containers. Smaller papers and polythene bags are equally used on small packages.

Figure 6-28: Sacks

Baskets: The use of baskets in packaging dated back thousands of years, in that they are cheap, light weight, easily repaired, durable and fairly strong. Baskets or wicker containers were used for the jobs we would now associate with corrugated cardboard cartons and plastic crates.

Figure 6-27: Baskets and hampers

Wicker (basket work) boxes were used for a wide range of goods including fruit and vegetables. Round baskets were used for shipping fruit; these were often tapered so they could be nested for shipping back when empty. They had woven handles near the top and were formed into a tall stack (perhaps seven feet long) with lines passed down through the handles. These were then stood on end in open wagons and roped in to secure them.

Sacks and bags: Canvas or 'sailcloth' is a weave not a material, actually 'twill' in which the threads running north-south are more numerous than those running east-west. Canvas is made from hemp or more recently cotton and cotton canvass. In agriculture sackcloth bags were used for storing and transporting bulk foods such as onions, potatoes, peanuts, Soya, sugar, salt, flour, rice and other grains and by farmers for holding seeds and chaff, animal foods and fertilizer etc.

Figure : Typical pre war grain sacks

Bags was a general term covering everything from small 1 lb (0.5 Kg) bags used for green beans to very large three or four bushel hessian sacks used for grain. Depending on the type of grain the larger sacks could weigh in at up to two and a half hundredweight when filled (the accepted weight when empty was 4lb or about 2Kg).

Cartoons: Cartoon containers can be constructed from hard paper, cardboard or particle/fiber boards, wood and wire. A special closing tool makes bending the wire loops on the crate's lid easier for packers to do. Wire-bound crates are used for many commodities including melons, beans, eggplant, greens, peppers, squash and citrus fruits.

Figure 6-27: A wire-bound wood crate

Cases and crates: Most items were shipped in light wooden crates and cases, often packed with straw. Cases and chests are technically closed wooden boxes and crates are open wooden frames forming a box shape. Crates can be made up using Slaters plastic strip,

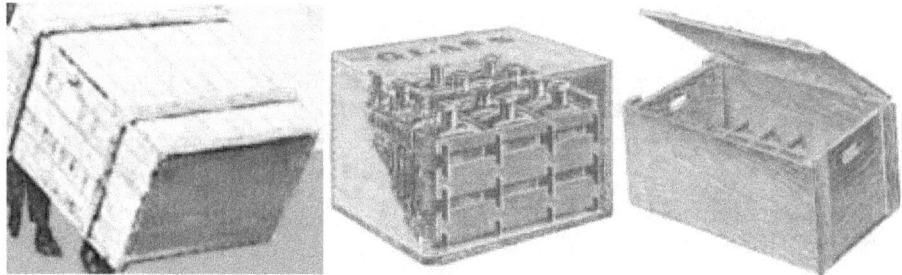

Figure 6-27: A wire-bound wood and bottle crates

Vegetables such as cabbages were shipped in light wooden crates; liquids can be shipped in containers made of glass, wood or metal.

Casks, barrels and kegs: A cask is a container made of wooden staves held together with hoops. Casks could be broken down into staves, ends and hoops for moving when it is empty. A 'barrel' is technically a single unit of casks with a capacity of 30-40 gallons (136-182 litres). The characteristic bulging shape of a barrel enable it to be rolled around easily and they were seldom seen on their ends.

Figure 6-27: Barrels

Kegs are small barrel shaped casks with a capacity of ten gallons or less (45.5 litres just under two foot long by just over a foot wide). The term is also used for the small steel cylindrical pressurized containers used for modern 'pasteurized' beer.

Tins and steel drums: Tins are thin sheet metal containers having a close fitting metal lid held in place by the glued-on paper wrapper laid round the sides which may or may not be tin-plated. One common application of tin was in the packaging of biscuit.

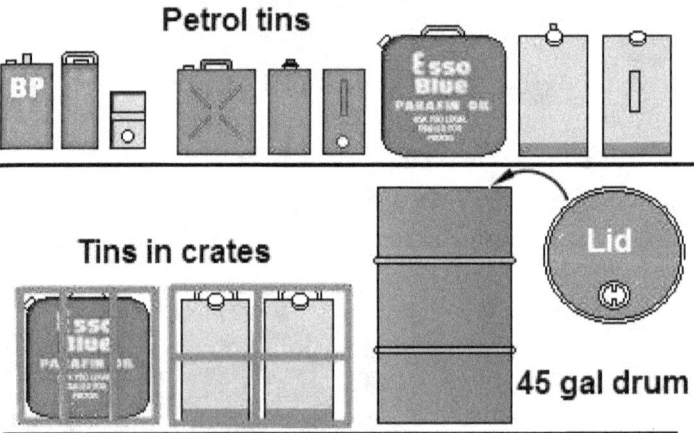

Figure 6-27: Barrels

Steel drums had ribs, either metal strip welded on or made by adding corrugations to the wall of the drum during manufacture. The corrugated drums were and are mainly used for chemicals with petroleum oils (mainly lubricants) coming a close second. Plain sided drums are usually associated with chemicals such as bleaching powder. The modern 'rolled steel drum' is made from a strip of 'cold rolled' steel 1-2mm thick (supplied as a stack of flat sheets or as a tightly rolled coil of sheet metal) into which ribs are pressed. Lengths of this are cut off and formed into a tube, the seam is welded and ends added.

6.7 Crop residue processing techniques

Kneading and cutting processes

Kneading is a new kind of straw processing technique combining chopping with grinding. The machine processes residues, especially maize stover, into thin thread segments of 8-19 cm, completely destroying its node structure.

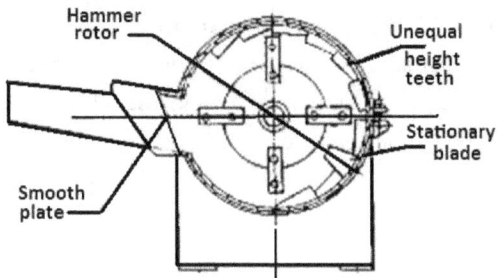

Figure 6-33: Illustration of kneading machine structure

Palatability is greatly improved, and the intake of the entire crop is also increased to 95 percent from the original 50 percent. The structure of the kneading machine is shown in (Figure 6-33). In operation, the rotor, diameter 40 cm, rotates at a high-speed (2856 rpm) driving 16 hammers arranged in 4 groups, which impact the straw fed continually. A tilted serrated plate, whose teeth are arranged helically and with changeable height, and 6 stationary blades are fixed to the concave plate of the machine, in order to keep the impacted straw moving in axially with the help of a fan.

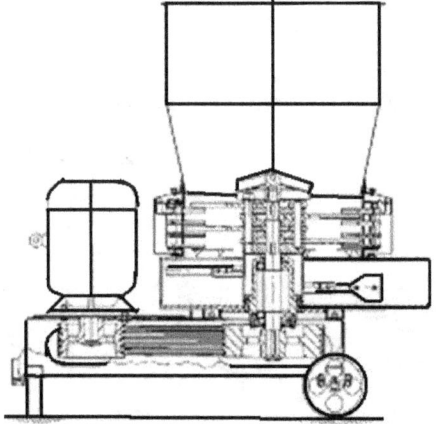

Figure 6-34: Type 9LRZ-80 kneading machine (*China Agricultural University*)

Kneading and cutting machines combine cutting, kneading and mixing in one operation. Maize stover, wheat and rice straw, bean vine and tuber vine can be processed to thread-like soft material. The future developments include the replacement of the cutting blades in the chopper by running blades; and the incorporation of a hammer in the kneading machine and of a mixing rotor in the mixer.

Straw choppers

The chopper is mainly used for stalk forage, such as rice straw, wheat straw, maize stover and maize for ensiling. Straw choppers can be classified by size into small, medium and large. The small-size chopper is mainly adapted for chopping dry straw or silage on small-scale farms. The large chopper - also called a silage chopper - is mainly used for silage on cattle farms. The medium chopper is normally suited to cutting dry straw and silage, so it is called a straw-silage chopper.

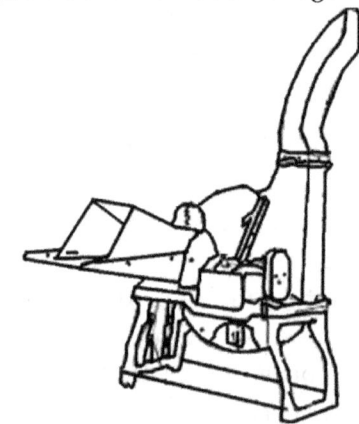

Figure 6-35: A typical straw chopper

Choppers can be divided into cylinder or flywheel types, according to the mode of cutting. Large- and medium-size choppers are generally flywheel types, to facilitate throwing silage, but the majority of small choppers are cylinder type.

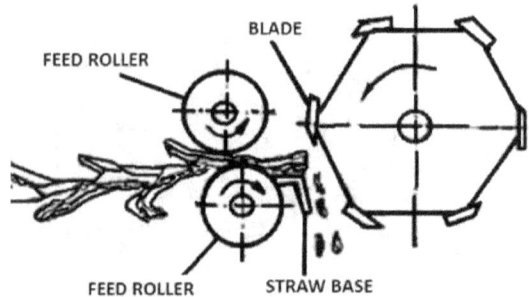

Figure 6-36: Chopping action of a cylinder type straw chopper

Large and medium choppers are usually equipped with road wheels for easy movement, while small-size choppers are normally stationary. The machine consists primarily of mechanisms for feeding (a chain conveyor, pressing rollers, and upper and lower feed rollers), chopping and throwing which consists of a main shaft, a blade rotor, rotating blades, a throwing vane and stationary blades, with a transmission, a clutch and a frame.

Flywheel choppers

Flywheel chopper operation is illustrated in Figure 6-37, showing a feed chain, upper and lower feed rollers, a stationary lower blade, a cutter and a throwing fan. The straw is fed via the feed chain into the feed rollers, pressed and moved forward by them, then cut into pieces by the combination of upper and lower blades, and it is finally blown by the fan to the storage site or silo.

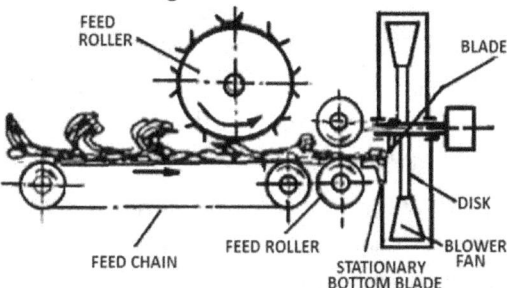

Figure 6-37: Chopping action of a disk chopper

Harvesting machine for maize silage

Stover can be harvested, chopped and loaded by a long-stalk cutting platform fitted to a silage harvester (also called a silage combine), and then put into the silo. The cylinder-type forage harvester is equipped with a long-stalk cutting platform and accomplishes picking, husking, ear collecting, chopping, and returning to the ground or throwing into a trailer as a single-pass operation.

Figure 6-38: Cylinder-type silage/forage harvester with thrower

Straw collector, loading and transport

Straw collectors are used to collect straws into small stacks, or to bring them to a large one. It consists of a collecting platform, left and right handspikes, a frame and a pulley support (Figure 6-39). The collecting platform includes a collecting fork, a side bar and a fence. The machine slides forward, with an angle of 5-7° from the ground, to collect straw with the fork. When the fork is full, hydraulic arms are engaged to lift it about 30 cm high from the ground.

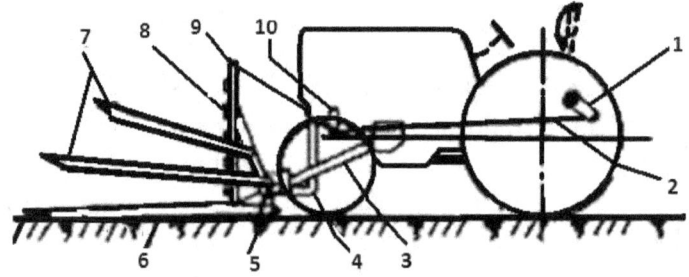

Figure 6-39: Straw collector

[1. Hydraulic arms; 2. Steel cable; 3. Handspike; 4. Frame; 5. Suspension spring; 6. Collecting fork; 7. Side bar; 8. Fence; 9. Supporting frame; 10. Pulley support]

Then the tractor moves straw to the stack. After lowering the platform, the tractor is backed to unload the straw. There is no stereotype for this machine. Because of its simple structure, users can construct them themselves.

Flail type straw chopper cum spreader

The machine consists of a rotary shaft mounted with blades known as flail to harvest straw and chopping unit consisting of knives. The straw after cutting by the flails, pass on to the chopping mechanism. The chopping mechanism chops the cut straw to 50 mm size and spreads it in the field uniformly. The cutting unit has 38 flails mounted in three rows. The chopping mechanism has 300 mm diameter cylinder with six rows of serrated knives and four counter rows each having 22 knives fixed at the bottom.

Shrub master

Shrub Master is a tractor PTO operated equipment. It consists of cutting blades (swinging flails) joined to the bar, gear box for transmission of power at right angle, universal joints with telescopic shaft to connect the tractor PTO and gearbox, adjustable side skids for controlling cutting height of shrubs or grass, safety guard and hitching frame. The bar having cutting blades at the ends is mounted on the gearbox shaft. Thus, the vertical shaft of the gearbox provides rotary motion to the bar. The cutting blades mounted on the bar swing open to the cutting position due to centrifugal force as the bar rotates in the horizontal plane. The cutting takes place purely through impact and flails need not be sharp-edged. The blades are made of medium carbon steel or alloy steel and hardened.

Straw stacker

There are many types of stacker, including derrick stackers, fan stackers, conveyor belt stackers, slide stackers and hydraulic stackers. For a hydraulic stacker (Figure 6-40), the collecting platform is lowered and the straw-pushing board is moved back by the hydraulic cylinder.

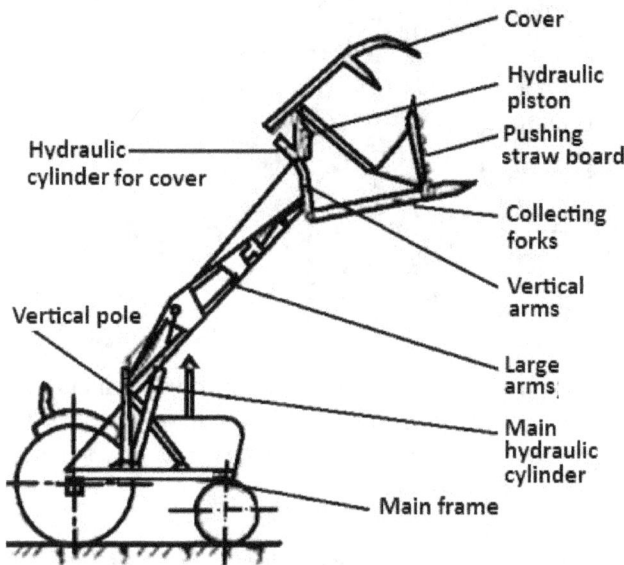

Figure 6-40: Straw hacker

The collecting fork is used to gather the straw until it is full while the large arm (6) lifts it. A hydraulic cylinder opens cover (1) and the hydraulic cylinder (2) moves the straw pushing board (3) forward and pushes the straw out.

Pick-up-and-press stacker

The flail type pick-up chopper is used to pick up and chop straw. This consists of a flail-type pick-up chopper, a blower tube with rectangular cross section, a directing cover, a movable top cover, a chamber, a chamber rear door, and a chain conveyor for discharging. The airflow, generated by the high-speed rotation of the chopper, blows straw into the chamber through the tube and directing cover. When the chamber is full, the tractor stops moving forward. The top cover is moved down by the compression mechanism to press the straw into the chamber, and then the top cover is lifted again. The machine continues to go forward picking up straw again.

Balers

A baler is a machine used to compress hay or straw into bales for easy transport and storage. A bale is the simplest minimum package for marketing. Balers are divided into stationary balers and field balers. They are further classified into rectangular balers and round balers according to the bale shape produced. According to density of bale, they could be high (200-350 kg/m³), medium (100-200 kg/m³) or low density (<100 kg/m³) balers.

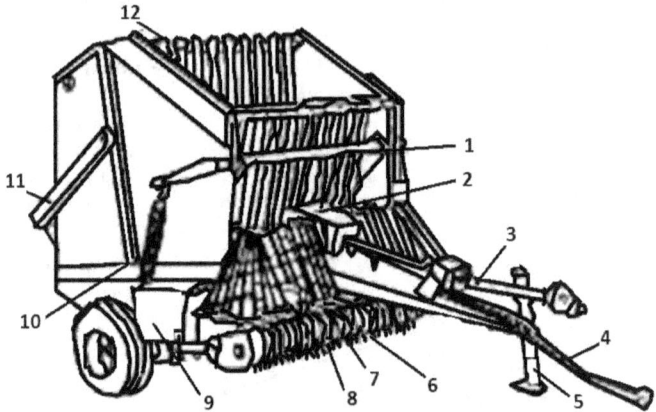

Figure 6-41: Structure and operation of balers

(a) *[1. Swing arms; 2. Gear box; 3. Gearing shaft; 4. Hydraulic power supply tube; 5. Support frame; 6. Pick-up reel; 7. Tube for the twine; 8. Twine cutter; 9. Twine box; 10. Tension spring; 11. Rear discharge door; 12. Belt of wrapping and pressing chamber.].*

Rectangular balers: The rectangular balers consists principally of a pick-up reel, a conveying and feeding system, a compressing chamber, a bale density adjuster, a bale length controller, a needle-and-tying mechanism, a crank-linkage mechanism, a power transmission, and hauling system. It is powered from the power take-off (PTO) of the hauling tractor. The straw windrow is lifted from the ground by a pick-up reel having spring teeth and transferred continuously to a conveying and feeding mechanism as the baler moves forward along the windrow. The conveying and feeding mechanism pushes individual charges of hay into the bale chamber from the side at intervals when the piston is withdrawn.

Figure 6-42: Rectangular balers

The piston reciprocates under the function of the crank-linkage mechanism to press the material into the bale. When the bale reaches the required length, the needle-and-tying mechanism is engaged automatically to bind the bale, which is then pushed out from the chamber by successive bales and is discharged to the ground.

Round baler: The wrapping action of pick-up bailer takes place in 3 stages: Forming of the core of the bale, *making* the bale and discharging the bale.

Figure 6-43: Pick-up round bailer

Bale wrapping

Wrapping of straw after baling is important for effective material handling. Bale wrapping materials are net wrap, baler twine and/or stretch film. The choice of which material to use plays a vital role in the quality of your work. High-quality products are crucial for the production of perfect round and square silage bales, whereas crop packaging of inferior quality may cause costly down time.

Figure 6-46: Wrapped square and round bale

Net wraps: Net wrap is made of polyethylene, derived from crude oil. Net wraps are made from highly tear-resistant warp threads woven into a patented pattern, which run lengthwise inside the net (Figure 6-47). The wefts threads are woven between these warp threads in a complex process. These weft threads, too, are high tensile and offer a generous reserve length not found in many ordinary net wraps.

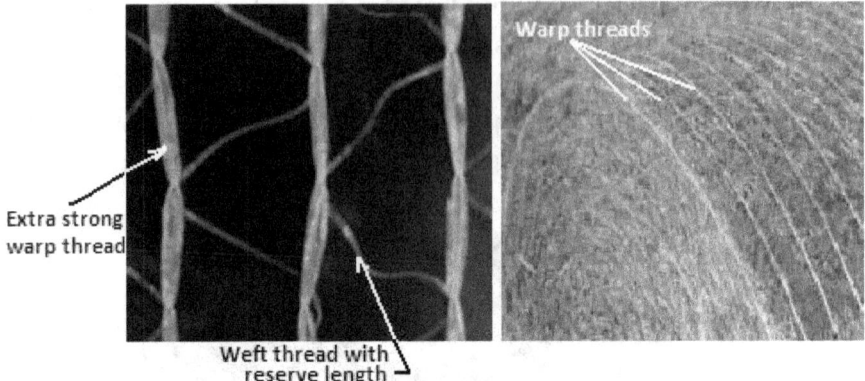

Figure 6-47: Warp and weft threads in net wrap

The presence of shoulder and net shrinking after wrapping are critical disadvantages of net wraps. 'Shoulders' on the bale's edges are a clear sign of a net wrap's poor edge to edge spreading ability. Shoulders are inevitable with poor quality net wrap, even though the baler is set to optimum performance. This is due to the weft threads inside the nets, which often lack the required reserve length. High quality net wraps have high resistance to tearing.

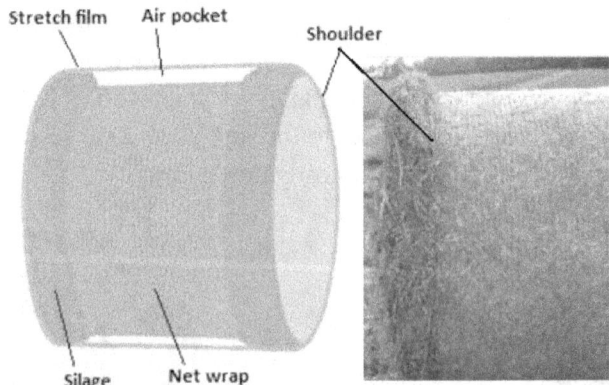

Figure 6-48: Bale wrap configuration showing shoulder

Baler twine: Baler twine are products based on synthetic polypropylene, which is derived from crude oil. These products have excellent quality for high resistance to tearing, high knot strength and provide extremely high bailing pressures in high capacity balers.

Figure 6-49: Baler twines

Twine properties

Knot strength: Knot strength expresses the knot's strength to resist force (expressed in kg/lbs) before it tears. On the baler, the twine is worked hardest in the knot, which is due to the knotting process. Therefore, this is the area that is most susceptible to tearing. Standard knot strength today is approximately 220 kg/ 485 lbs (KRONE excellence).

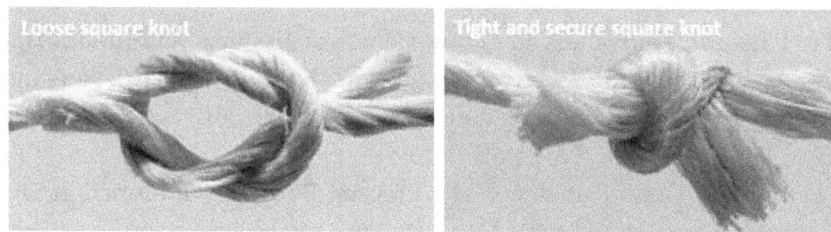

Figure 6-50: Baler twine knots style

Resistance to tearing: The higher the claimed resistance to tearing, the higher the overall twine stability and strength. Ordinary twines with a low resistance to tearing are often unable to resist high bale densities.

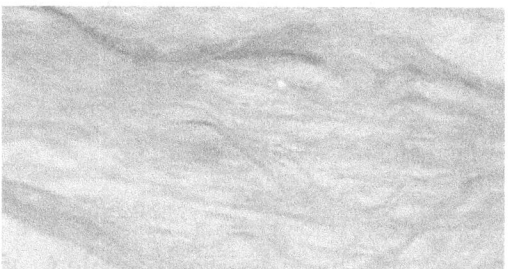

Figure 6-51: Twine strands

Optimum fibrillation ensures tight knots: Fibrillation is a separate step within the production process, where the twine is spun from film tape after this received special treatment. As you untwist the twine, you see that the strands are actually treated film tape. Best possible fibrillation produces very robust twine and tight knotting.

Ideal spooling: Baler twine should unroll easily and uniformly. Daily checks on the spooling process assure *Krone* excellent twine performs absolutely trouble-free inside the baler.

Further reading

Agricultural Engineering International: the CIGR Journal of Scientific Research and Development. Manuscript FP 03 013. Vol. VI. August, 2004.

Beneficial Management Practices, (2004): Environmental Manual for Alberta Cow/Calf Producers Alberta Agriculture, Food and Rural Development (AAFRD)

Bilanski W.K., Collins, S.H. Chu. (1962) Aerodynamic properties of seed grains. Their behavior in free fall Jour. of Agricultural Engineering 216-219.

Bill Mecke, P.E., 2011. Advances in Screw Conveyor Systems KWS Manufacturing Company Ltd. http://www.kwsmfg.com/about/news/article/advances-in-screw-conveyor-systems.htm. date accessed: 15/09/2012

Cherney J.H. (2006). Grass Pelleting – The Process, Bioenergy Information Sheet #7 Dept. of Crop & Soil Sciences http://www.GrassBioenergy.org

David A. Fairfield, (2003). Pelleting for Profit - Part 1. Feed and feeding Digest. Volume 54, Number 6, November 13, 2003. *National grain and feed association.*

György Sitkei (1986). Mechanics of Agricultural Materials. Elsevier Science Publishers, Amsterdam, The Netherlands.

Handbook of Crushing With Posimetric Feeders (2006). Pennsylvania Crusher www.penncrusher.com

Hooper A.W., Harries G.O., Ambler B. (1976). A photoelectric sensor for distinguishing between plant material and soil. Jour. of agric. Eng. Res. Vol. 21, 67-75.

Igbeka J.C and Oluleye, A. (1986). Some ergonomic consideration in operating a pedal thresher, *Agricultural journal vol. 5, No.1 p 87*

Janet Bachmann (2004).Oilseed processing for small-scale producers ATTRA -NCAT Agriculture Specialist Revised May

Janet Bachmann and Richard Earles (2000) Postharvest Handling of Fruits And Vegetables NCAT Agriculture Specialists August 2000

Jeffrey H. McCormack,(2004). Seed processing and storage Principles and practices of seed harvesting, processing, and storage: an organic seed production manual for seed growers in the Mid-Atlantic and Southern U.S.

Lisa Kitinoja and Adel A. Kader, 2002. Small Scale Postharvest Handling Practices: A Manual for Horticultural Crops (4th edition).

Mike Foale, 2012. Cocosplit – The Simple Way to Neatly Split Your Coconut. *http://www.cocosplit.com/action.html. date assessed 17/09/2012*

Munoz-Hernandez G., Gonzalez-Valadez M., and Dominguez-Dominguez J. (2004). "An Easy Way to Determine the Working Parameters of the Mechanical Densification Process".

Nelson, S. O. (1980). Moisture-dependent kernel and bulk density for wheat and corn. Trans. ASAE 23:139-143.

Norder, R., and S. Weiss. 1984. Bucket elevator design for farm grains. Paper No. 84-3512. St. Joseph, MI: ASAE.

Nuri N. Mohsenin. (1972). Mechanical properties of fruits and vegetables review of a decade of research Applications and Future needs. Trans. Of the ASAE.

Oloko S. and Agbetoye L. (2006). "Development and Performance Evaluation of a Melon Depodding Machine". Agricultural Engineering International: the CIGR Ejournal. Manuscript PM 06 018. Vol. VIII. August, 2006

Raji, A. O. and Akaaimo D. I. (2006) Development and evaluation of a threshing machine for *prosopis africana* seed Web Page

PHTRC, 1984. A portable sorting table. Appropriate Postharvest Technology 1(1):1-3. (Post-Harvest Training and Research Center, Department of Horticulture, University of the Philippines at Los Banos.)

Reid W.S., Buckley D.J., Mason W. (1976) .A photoelectric seed counting Detector Vol. 21, 213-215. Jour. of agric. Eng. Res.

Reznicek R. (1988) Physical properties of Agricultural materials and their influence on Design and performance of Agricultural machines and technologies. Opening speech of the third international PPAM conference, Prague, Czechoslavakia. Hampshire Publisher USA.

Richard H. Leaver, P.E, (2003). Wood pellet fuel and the residential market. Andritz, Inc., Sprout Matador Division, 35 Sherman Street, Muncy, PA 17756

Tates W.E., Stephenson J., K. Lee, Akeson N.B. (1973). Dispersal of Granular Materials from Agricultural Aircraft. Trans. of the ASAE 609-314.

Schell, T.C., and E. van Heugten, 1998. The Effect of Pellet Quality on Grower Pigs. Journal of Animal Science, volume 76 (supplement 1):185. American Society of Animal Science, Savoy, Ill.

Whitney R.W, Porterfield J.G. (1968). Particle separation in a pneumatic covering system. Trans. of the ASAE Vol 11 No4 477-479

Zatari, I.M., P. R. Ferket, and S. E. Scheideler, 1990. Effect of pellet integrity, calcium lignosulfonate, and dietary energy on the performance of summer raised broiler chickens. Poultry Science. Volume 69:198. Poultry Science Association Inc., Savoy, Ill.

CHAPTER 7

Fruits Processing and Storage

7.   Introduction

Fruit and vegetable are agricultural products that form essential part of our balanced diet however with high degree of perishability. It is becoming increasingly important to present top-quality products to discerning consumers, locally and internationally.

Postharvest technologies need to take account of the living nature of horticultural products, and in particular their susceptibility to physical, pathological, and physiological deterioration. Most major supermarket chains buy horticultural products to specification (size, weight, colour, and freedom from defects); buyers are rejecting products that deviate from pre-contracted specification ranges.

Both physical and temperature abuse of agricultural products can lead to increased decay. Optimum storage conditions (such as optimized refrigeration and controlled-atmosphere storage) can reduce respiration, thereby achieving slower deterioration and enhanced storage life. Approaches such as preventing ethylene contamination of the storage environment from external sources or ethylene producing crops, combined with new engineering and molecular approaches will be important in the production and handling systems used to get products to market and maximize storage life.

It is therefore imperative that growers, exporters, shippers, and scientists work together to ensure that customer quality standards are being met. An important first step in this process is to ensure that all concerned understand the nature of product perishability and appreciate important biological factors that influence deterioration.

7.1   Fruits and vegetable

Fruits are seed-bearing structures or parts of a flowering plant. A fruit is actually a ripened ovary, a component of the flower's female reproductive structure. Fertilization of the egg, or female sex cell, within the ovary stimulates the ovary to ripen, or mature. Depending on the type of plant, the mature ovary may form a juicy, fleshy fruit, such as a peach, mango, apple, plum, or blueberry. Or it may develop into a dry fruit, such as an acorn, chestnut, or almond. Grains of wheat, corn, or rice also are considered dry fruits. Certain foods commonly termed vegetables, including tomatoes, squash, peppers, and eggplant, technically are fruits because they develop from the ovary of a flower.

7.2   Importance of fruits and vegetables

Fruit and vegetables are vital to humans and animals alike and as such has the following importance:

1.   They form essential part of human balanced diet.
2.   Many species of mammals, birds, and insects rely on fruit as diet.
3.   Fruits also play a critical role in dispersing seeds, increasing the likelihood that at least some will land in an environment favorable for germination, or sprouting, which helps to propagate the plant species.
4.   Fruit and vegetables serve as important sources of digestible carbohydrates, minerals, and vitamins, particularly vitamins A and C.
5.   In addition they provide roughages (indigestible carbohydrates), needed for normal healthy digestion

7.3   Horticultural products as living entities

Freshly harvested horticultural products remain alive after harvest, in contrast with other food products such as cereals, meat, and milk-based items. Post harvest technologists have attempted to maintain quality by slowing deterioration to improve storage and shelf life, and thus ensuring that consumers have high quality fruits and vegetables to purchase. They seek to control the handling, transport, and storage conditions to ensure optimal quality.

After harvest, horticultural products do not remain in a constant condition but continue to develop through the following processes that are genetically predetermined:

Process 1: Maturation
Process 2: Ripening
Process 3: Senescence
Process 4: Death.

It is worthy of note that the developmental processes of *maturation* (maturity) and *ripeness* merge and overlap. Each of these developmental processes is discussed below.

7.3.1 Fruit maturity

Maturity is a stage in fruit development when all the physiological processes have fully developed to its peak. At this stage, the fibrous and the non fibrous materials constituting the fruit has reached its peak. When harvest at this stage, the fruit can be processed into desired product or kept for distance transport. The maturity stage at harvest is paramount in determining the final product quality after postharvest ripening.

The word mature is derived from Latin word *'Maturus'* which means ripen. It is that stage of fruit development, which ensures attainment of maximum edible quality at the completion of ripening process.

The level of maturity actually helps in selection of storage methods, estimation of shelf life, selection of processing operations for value addition etc. Maturity has been divided into three categories i.e. physiological maturity horticultural maturity and commercial maturity.

1. *Physiological maturity*: Physiological maturity refers to that stage of development when maximum growth has occurred and proper completion of subsequent ripening can occur even if the product has been harvested. This is the stage when a fruit is capable of further development or ripening when it is harvested i.e. ready for eating or processing.
2. *Horticultural maturity*: This refers to the stage of development when plant and plant part possesses the pre-requisites qualities for use by consumers for a particular purpose i.e. ready for harvest.
3. *Commercial maturity* is that stage of development of a fruit or vegetable that is required by the market (retailer or consumer); it may have little relation to physiological maturity and may occur at various stages of ripeness depending on individual consumer preference.

Maturity index

Not all the fruit in an orchard, or even on the same tree, will ripen and turn colour at the exact same time, so a maturity index was developed to help producers numerically categorize their fruit's maturity level. A maturity index number allows producers to evaluate their varieties under their own specific growing conditions over a number of years.

Importance of maturity indices

Maturity index is useful in achieving one of the following objectives

- Ensure sensory quality (flavour, colour, aroma, texture) and nutritional quality.
- Ensure an adequate postharvest shelf life.
- Facilitate scheduling of harvest and packing operations.
- Facilitate marketing over the phone or through internet.

Determination of maturity index

Maturity index can be evaluated by one of the following methods:

1. *Determination of maturity index by colour changes*

The loss of green colour of many fruits is a valuable guide to maturity. There is initially a gradual loss in intensity of colour from deep green to lighter green and with many commodities, a complete loss of green colour with the development of yellow, red or purple pigments.

Ground colour as measured by colour charts, is useful index of maturity for apple, pear and stone fruits, but is not entirely reliable as it is influenced by factors other than maturity. For some fruits, as they mature on the tree, development of blush colour, that is additional colour superimposed on the ground colour, can be a good indicator of maturity. Examples are red or red-streaked apple cultivars and red blush on some cultivars of peach.

Objective measurement of colour is possible using a variety of reflectance or light transmittance spectrophotometer. Colour perception depends on the type and intensity of light, chemical and physical characteristics of the commodity, and person's ability to characterize colour. Although human eye is used to evaluate colour, however, results can vary considerably due to human differences in colour perception.

Therefore, an instrument (objective method) is used to provide a specific colour value based on the amount of light reflected off the commodity surface or light transmitted through the commodity. Maturity Index (MI) is determined as follows:

a. Take a random sample of about 2kg of fruit from several trees (high and low) of the same variety in the area where harvest is imminent. Collect all the fruit from a branch here and there rather than individual fruits.

b. Randomly select 100 fruit out of the sample bucket. Repeat 3-10 times until all the fruit is gone.

c. Separate each 100 fruit sample into eight colour categories based on the colour chart below:

0 = Skin colour is deep green-fruit is still hard

1 = Skin colour is yellow-green -fruit starting to soften

2 = Skin with less than half the fruit surface turning red, purple, or black

3 = Skin colour with greater than half the surface turning red, purple, or black

4 = Skin colour all purple or black with all white or green flesh

5 = Skin colour all purple or black with less than half the flesh turning purple

6 = Skin colour all purple or black with greater than half the flesh turning purple

7 = Skin colour all purple or black with all the flesh purple to the pit

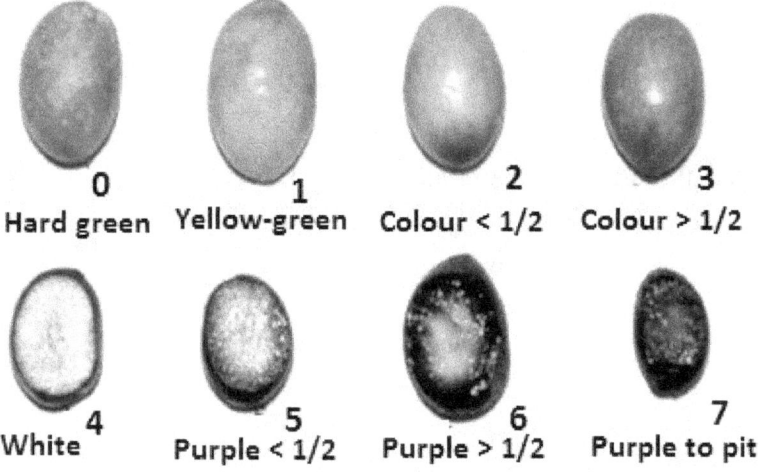

Figure 7- 1: Fruit colour chart

d. Multiply the number of fruits in each colour category by the number of that colour category (0 to 7)

e. Add all the numbers together and divide by 100:

$$MI = \frac{Ax0 + Bx1 + Cx2 + Dx3 + Ex4 + Ex5 + Gx6 + Hx7}{100} \quad \text{............6.2}$$

Letters (A-H) represents number of fruit in each category.

For example, the separation of the fruits in tray on the left of Figure 7- is shown in the trays to the right. The fruits were separated as follows

Figure 7- 2: Determining maturity index by shape of shoulder

A: 5 green x 0 = 0 ,

B: 20 yellow − green x 1 = 20 ,

C: $20 < \frac{1}{2}$ colour turning red, purple or black x 2 = 40 ,

D: $28 < \frac{1}{2}$ colour turning red, purple or black x 3 = 84 ,

E: 12 black/white flesh x 4 = 48 ,

F: 8 black $> \frac{1}{2}$ purple flesh x 5 = 30 ,

G: 5 Black $> \frac{1}{2}$ purple flesh x 6 = 30 ,

H: 2 black flesh to pit x 7 = 14 ,

$$MI = 0 + 20 + 40 + 84 + 48 + 30 + 14 = 176 = \frac{276}{100} = 2.76$$

2. *Determination of maturity index by size and shape*

Maturity of fruits can be assessed by their final shape and size at the time of harvest. Fruit shape may be used in some instances to decide maturity. For example, the fullness of cheeks adjacent to pedicel may be used as a guide to maturity of mango and some stone fruits (Figure 7-3).

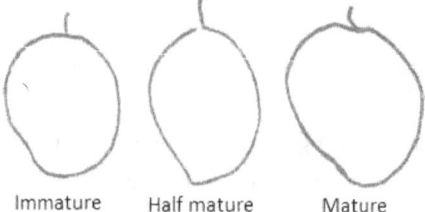

Immature   Half mature   Mature

Figure 7- 3: Judging mango harvest maturity by shape of shoulder

3. Determination of maturity index by firmness test

As fruit mature and ripen they soften by dissolution of the middle lamella of the cell walls. The degree of firmness can be estimated subjectively by finger or thumb pressure, but more precise objective measurement is possible with pressure tester or penetrometer

7.3.2 Ripeness

This is not a physiological process in fruit development but a declining stage in the life of the fruit. Ripening is thus the change of the green chlorophyll content of the fruit into various colourations as an indication of quality degeneration. Thus ripening is a state of degeneration of fruit through change of colour, texture, firmness, sugar content, acidity content etc.

The word ripe was derived from a Saxon word '*Ripi*', which means gather or reap. This is the condition of maximum edible quality attained by fruit following harvest. Only fruit which becomes mature before harvest can become ripe. When a product has attained the state or stage of maturity, it is said to be mature and ready for ripening.

Ripening

Final eating quality is critically dependent on harvesting at the correct maturity stage, so that normal ripening can occur with the concomitant development of flavour, texture, aroma and juiciness required by consumers. In many fruits, ripening occurs either on or off the tree. Optimum eating quality for many vegetable crops is attained before full maturity. Examples include peas, green beans, sweet corn, asparagus, and leafy vegetables; if these products are left attached to the parent plant and not harvested at the correct time, their quality is much reduced.

Ripening involves a series of changes occurring during the early stages of senescence of fruits in which structure and composition of unripe fruit is so altered that it becomes acceptable to eat. The developmental processes of maturation (Maturity) and ripeness merge and overlap.

Ripening is a complex physiological process resulting in softening, colouring, sweetening and increases in aroma compounds so that ripening fruits are ready to eat or process. The associated physiological or biochemical changes are increased rate of respiration and ethylene production, loss of chlorophyll and continued expansion of cells and conversion of complex metabolites into simple molecules.

When fruits on a tree grow to its intended size and shape – *maturity* then, within a week or so, it *ripens* with visible changes in the following qualities:

1. *Aroma*: Bitter and astringent phenols fade away (their job was to discourage animals before the seed was ready), and nice aromas are produced (to encourage animals). This normally only happens while fruit still attached to tree. Ethylene gas is the "ripening hormone" that coordinates the ripening.
2. *Sweetness*, in the form of sucrose or fructose. It can come as sweet sap while attached to the tree or in some fruits by converting the fruit's stores of starch/glucose/acid.
3. *Juiciness and softness*: An enzyme polygalacturonase attacks pectin in the cell walls making cells slide around (softness) and spill their contents (juiciness). Acids are used up in this, making the fruit less sour.
4. *Colour changes*: The colour brightens, forms a waxy sheen to slow loss of water. Look at the background colour, not the red blush which growers have bred so it appears even before ripeness.

7.3.3 Senescence

Senescence can be defined as the final phase in the ontogeny of the plant organ during which a series of essentially irreversible events occur which ultimately leads to cellular breakdown and death.

7.4 Fruit and vegetable storage conditions

When buying fruit samples from loose fruit rack or open market, smell it and check for no bruising or mould or musty damp. Handle ripen fruit carefully, since ripe fruit spoils easily. Don't refrigerate a fruit before it ripens, since this ruins ripening. Once ripe, or immediately for non-ripening fruit, it is advisable to eat the fruit or otherwise,

1. Store in refrigerator to slow down respiration,
2. Store in plastic bag to stop moisture loss, or
3. Unsealed to avoid fermenting.

Deterioration of fresh products

Deterioration commences at harvest; post harvest technologies are designed to slow the rate of ripening and senescence and hence quality decline. If deterioration is rapid, poor-quality product can be removed at the point of production or packing at which quality inspection occurs; if deterioration is slow the product may pass initial quality inspection yet be of reduced acceptability to consumers because of poor appearance, texture, or taste.

Deterioration results from three main types of effects: physical, physiological, and Pathological.

1. Physical effects

Agricultural products can sustain mechanical or physical damage at all stages of the chain from harvest to consumption. Bruises, cuts, abrasions, and fractures occur as a result of poor handling or inadequate packaging. Such damage dramatically increases water loss and susceptibility to infection by post harvest fungi and bacteria. Product firmness and water status influence susceptibility to mechanical damage. Development of gentle yet effective handling systems and appropriate packages, together with education of personnel, are required to minimize physical damage.

2. Physiological effects

All living things respire to generate energy for continued metabolism. A simplified summary equation for respiration is

$$C_6H_{12}O_6 + 6O_2 \rightarrow 6CO_2 + 6H_2O + Heat\ energy \uparrow$$

Respiration rate of horticultural products varies, but as a general rule perishability is a function of respiration rate; the greater the respiration rate the more perishable the product and the shorter the time it can be stored and still maintain acceptable quality. Actively growing products, such as asparagus, or tropical fruit, such as mango, have high rates of respiration and limited storage life.

Respiration is highly temperature dependent; the lower the temperature (down to 0°C) of harvested fruit and vegetables the lower the respiration rate. Consequences of lowering respiration rate include

1. Reduction in carbohydrate loss
2. Decreased rate of deterioration

3. Increased storage and shelf life

3. *Pathological effects*

Any physical damage to produce provides an ideal entry point for pathogens. A wide range of fungi and bacteria contribute to post harvest losses in fruit and vegetables throughout the world, especially in situations in which cool-chain management is inadequate. These post harvest pathogens may infect produce at various preharvest stages through plant or fruit development, at harvest, or after harvest while products are in store or in transit to market. Removal of previously infected fruit and application of appropriate fungicides at bloom are the recommended means of control.

Methods of maintaining quality and extending life of harvested products

1. *Low-temperature storage*

Low-temperature storage is the major weapon that post harvest operators have to maintain quality and extend life of harvested products. Low temperatures not only reduce respiration rate, but also reduce water loss through transpiration and nutritional loss. Immediately after harvest, products should be placed in a well-ventilated shade environment to prevent large temperature increases that occur if products are exposed to direct sunlight.

Minimizing temperature variation from the optimum recommended during storage improves uniformity in product quality. Refinements are being made in refrigeration control systems, cool-store design, and pallet-stacking patterns to optimize air temperature and airflow through and around pallets in cool stores .This prevents hot spots from developing in localized areas within the store that would lead to higher respiration rates, more rapid deterioration, and hence poorer quality than in parts of the store in which product temperature was optimized.

2. *Rapid cooling*

Rapid cooling as soon as possible after harvest (to reduce respiration rate) is used commonly in the horticultural industry, particularly for very perishable products such as strawberries and apples. Hydro cooling, vacuum cooling, or forced air-cooling is used widely to remove field heat rapidly. The method chosen depends on the crop.

3. Packaging

Packages should be designed to allow cool air to flow directly over products, ensuring rapid temperature pull-down and consequent temperature maintenance. Regrettably, too many packages in current use are largely in penetrable to air movement, resulting in a slow temperature decrease, influencing quality negatively.

4. Controlled atmosphere

Controlled atmosphere storage has been the mainstay of the European and North American apple industries for many decades but has not been commercially successful for many horticultural crops. Adoption of this technology has been slow in other countries, although information on desired atmospheres for a range of crops is available.

5. Modified atmospheres

Modified atmospheres, in which the atmosphere is passively regulated depending on product mass, temperature, and nature of a polymeric film package, are being used increasingly for minimally processed or "fresh-cut" products. In addition, they can be used to reduce disorders in whole fruit.

Pre-harvest factors affecting postharvest quality

Ultimate eating quality of fruit is largely predetermined at harvest. Choice of cultivar, rootstock, irrigation, fertilizer regime, training, and pruning systems will impact fruit size and colour at harvest. Great care must be taken during harvesting of perishable fruits and vegetables to avoid physical damage. Any mechanical damage that occurs at harvest, during the movement of product to the pack house, or through grading and packing lines will result in enhanced respiration, elevated ethylene production, water loss and increased susceptibility to infection by postharvest pathogens, all of which can induce rapid deterioration and loss of quality.

A number of simple but effective steps can be taken to reduce physical damage from occurring during this phase of the harvesting and handling system .These include careful handling of the product at all stages of the operation, good sanitation and hygiene with all equipment maintenance of packing equipment to prevent excessive drops onto hard surfaces, and padding of all machinery surfaces on which products may impact. Curing is a simple method of reducing losses due to decay and water loss during storage of some products including some citrus fruit, root, tuber, and bulb crops.

7.5 Fruit harvest and handling methods

Fruit harvest methods

Fruits and vegetables are popularly harvested by two methods; hand picking and mechanical methods. In some countries fruit is harvested by hand, placed onto straw on the ground under the trees, hand-sorted, and packed into containers before leaving the orchard. Mechanical aids are available for harvesting in the form of gantries, picking ladders, and in some cases mobile conveyor systems, but more commonly the picker collects the fruit into a small holder such as a picking apron or bucket.

a. *Hand picking method:* In some countries fruit is harvested by hand, placed onto straw on the ground under the trees, hand-sorted, and packed into containers before leaving the orchard. Under these conditions consistent quality control is difficult to achieve among orchards, but fruit may sustain less handling damage. Fruit are easily damaged at harvest time, so care is required. Product for the fresh-fruit market mostly is harvested by hand into suitable containers.

b. *Mechanical method:* Various mechanical means have been developed for harvesting of fruits and vegetables. Common among them are tree shaking, mechanical aids and mechanical harvesters.

   i. *Tree shaking:* Harvesting involves shaking the tree or cane by mechanical vibration and catching the detached fruit underneath in a large blanket or net. However, these systems can cause significant damage to the crop and are generally only suitable for fruit to be used for processing. There are difficulties if the fruit on the tree do not all ripen at the same time.

   ii. *Mechanical aids:* Mechanical aids are improvements on the tree shaking and hand picking methods of harvesting fruits. Mechanical aids are available for harvesting in the form of gantries, picking ladders, and in some cases mobile conveyor systems, but more commonly the picker collects the fruit into a small holder such as a picking apron or bucket, which holds not more than 15 kg of fruit. Mangoes, papaya, apples, and fruit grown in tall trees can be harvested using picking poles. The fruit is separated from the tree by a sharp cutting edge on the end of the pole and falls into a net just under the cutter.

   iii. *Mechanical harvesters:* Mechanical harvesters have been developed for many crops including apples, strawberries, blackcurrants, blueberries, cherries, and raspberries. Some fresh vegetables are harvested by mechanical means. These include peas, beans, tomatoes, sprouts, and root crops, particularly if the product is for processing.

Harvest handling

Quality cannot be improved after harvest, but can only be maintained; therefore it is important to harvest fruits and vegetables at the proper stage and size and at peak quality. Fruit are easily damaged at harvest time, so care is required. Harvest should be completed during the coolest time of the day, which is usually in the early morning and late evenings and produce should be kept shaded in the field.

Crops destined for storage should be free as possible from skin break, bruises, spots, rots, decay, and other deterioration. Bruises and other mechanical damage do not only affect appearance, but provide entrance to decay organisms as well.

Postharvest handling of fruits and vegetable

Good product quality is attributable to appropriate production practices, careful harvesting, and proper packaging, storage, and transportation system. Handling is the final stage in the process of producing high quality fresh product. Being able to maintain a level of freshness from the field to the dinner table presents many challenges. Production practices have a tremendous effect on the quality of fruits and vegetables at harvest and on postharvest quality and shelf life.

Fruit and vegetables must be transferred from the field to the table, to arrive in a state that is acceptable to the consumer. Most fruit and vegetable crops begin to deteriorate as soon as they are harvested, and most are particularly prone to handling damage at all times. In general, the level of susceptibility of these products to handling damage is greatly underestimated, usually because the effects of mishandling do not appear until sometime after the damage occurred.

Poor handling and storage can easily result in a total crop loss anywhere. Not all losses are due to poor handling, but handling damage is known to accelerate other types of deterioration, particularly the development of molds and rots.

7.6 Postharvest operations

Clearly postharvest operation processes differ among crops, and some crops do not require all the stages listed or may require further operations to enhance the quality of the product. Major steps involved are as follows:

Transfer to field bin

The fruits harvested were transport to packing shed over rough farm tracks and other roads to the grading shed or market. This movement causes potential damage to fruit as there is no packaging to protect them. To prevent such damage, fruits can be transported in self-propelled field pack systems or crops collected into larger containers (field bins) which are transported out of the orchard.

Figure 7-4: Self-propelled field pack system

Care should be taken as the transfer into the field bin is a serious potential cause of damage, unless the pickers are well trained. Fruit-on-fruit impact and impacts against the sides and base of the bin are potential sources of severe bruising.

Unloading from field bin or pack

This operation needs to be achieved with minimal damage to products. Floating out fruits into channels using water works well for fruit with a density less than that of water. Apples and pears can be removed in the same way, provided a suitable chemical is added to the water to increase its density. If water cannot be used, then options include tipping the bin over slowly or using side openings on the bin. Submerging apples in the field bin in water results in little damage.

Figure 7-5: Unloading fruits from field bin at reception

Reception

At the reception, documentations on where and who the product is from, the harvest date and the delivery date to the shed, as well as full details about the crop and pesticides used during its growth of the crop arriving at a grading shed is recorded. This allows the grower to receive appropriate credit for the product and ensures that quality guidelines have been adhered to prior to arrival of the product at the shed. Reliable documentation requires careful inventory control, including clear procedures for recording all shipments and marking the field bins.

Figure 7-6: Fruit receptacle-bin at reception

Precooling

If the product cannot be dealt with immediately when it arrives at the packing shed, it is essential to minimize the deterioration of product. For most crops this involves reducing the temperature. Highly perishable leafy vegetable crops can be cooled rapidly in the field using mobile vacuum coolers

Figure 7-7: Single-bin drenching with a postharvest fungicide

Drenching

Drenching is done to reduce physiological disorders or infestations in harvested products. This can be done using an overhead spray system that washes through the field bin, or dipping the bin in a bath, which may be outside the grading shed.

Figure 7-8: Single-bin drenching with a postharvest fungicide

Washing

Field dirt must be removed before sorting. Providing fruit is not seriously soiled, the water dump followed by a series of rotating brushes will remove dirt, without causing damage to the fruit. However, there is some concern that brushing can increase water loss from some products such as citrus. Also, the surface of some fruits (especially pears) is easily damaged by brushing or rubbing. After washing, surplus water should be removed. This usually is achieved as the fruit passes over the brushes or absorbent foam rollers but may require forced warm air from fans.

Presorting

Presorting is used to remove fruit that clearly does not meet specified quality standards. This process helps to maximize throughput in the rest of the grading system. *Machine vision color sorters* remove poor color fruit, and *presizers* remove excessively small or large fruit, before manual grading fruit according to quality standards.

Figure 7-9: Presorting

Sorting (grading)

During sorting, fruit are graded (sorted) into categories. In many cases there are only two or three grades (e.g., export, local market, or juicing). Although computer vision systems are making significant advances, grading still is best achieved in most cases by human inspection. Sorters require suitable facilities to enable them to see defects clearly. Fruit must be rotated so that the sorters can see the entire fruit surface. Sorters will perform best if they work under good lighting conditions and in comfortable working positions.

Figure 7-10: Apple sorter

Singulating

If fruit are to be sorted according to weight, the singulator separates fruit into pockets or cups so that each fruit can be weighed independently. The fruit can then be separated into appropriate sizes by sorting according to the weight recorded for the cup. Various devices including transfer wheels and expanding belts can be used for singulation.

Figure 7-11: Sinclair multi-head singulator

Size band sorting

Size uniformity and conformity is particularly desirable for the purposes of fruit packaging and display. Some fruits and vegetable have a consistent shape, thus they can be conveniently weighed and sorted. The result is a product that is consistent in volume and shape and packs easily.

Figure 7-12: Band sorter

Electronic or mechanical methods can be used for weighing; shape sorters can be classified as mechanical or visual.

Mechanical weighing

In a mechanical system, each fruit is "weighed" as it passes each exit point. A mechanical lever releases the fruit if the weight exceeds the preset trigger level at that point. Drop points therefore must be arranged along the line in increasing weight, which can cause logistical problems. Mechanical systems are usually less costly and can be maintained and repaired by a trained mechanic.

Figure 7-13: Circular weight grader with shallow bins

Electronic weighing

Fruits can be weighed with either mechanical weighing scales or electronic weight sorters. In mechanical weighing systems, a quantity of fruit is arranged in the weighing tray and the resistance to weighing spring monitored by the needle and read on the graduation is recorded.

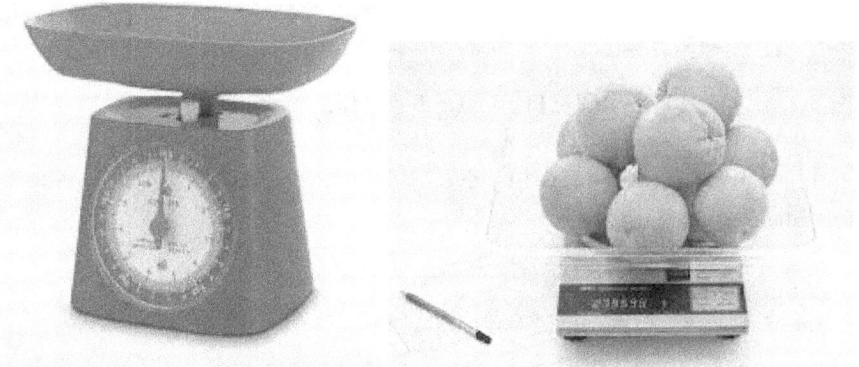

Figure 7-14: Weighing scales

In the electronic system all fruit are weighed at one point. Data is fed to a computer, which selects and preprograms the drop point. Fruit may rest on several fingers (rather than a single cup), which are programmed to release the fruit at the required point, and the fruit rolls off sideways.

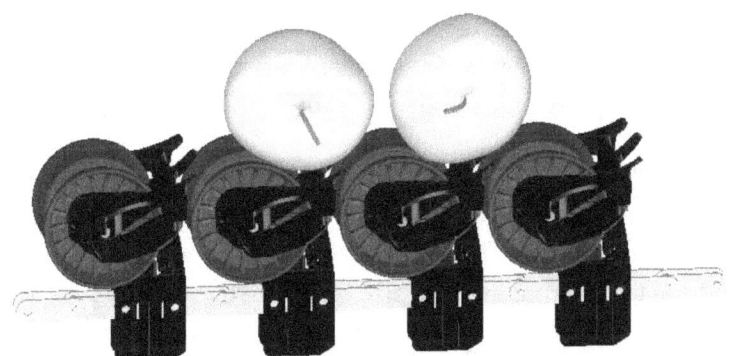

Figure 7-15: Electronic weight sorter

Size sorting

Some crops (e.g., tomatoes, carrots) produce inconsistent results if sorted by weight. Better results are obtained if they are sorted according to shape, using expanding screens or tapers. Fruit moves through the sorter and passes over or rolls along a continuously enlarging exit orifice until it is able to fall through into the appropriate container, such as in Figure 7-16.

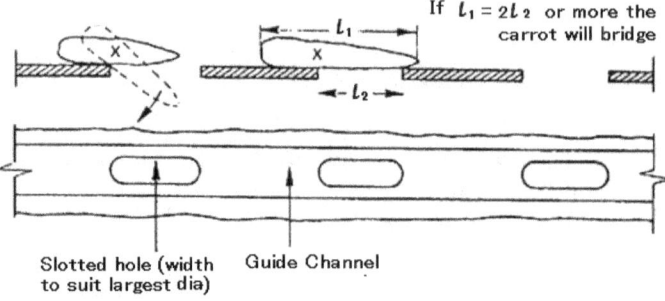

Figure 7-16: Fruit length grading mechanisms

In the field, crops also can be sized by eye, or by using a set of rings or plastic cards of predetermined size.

Image analysis sorting

Increasingly, image analysis is used to size fruit and vegetables. Cameras take a view of the product, and according to the algorithm the fruit is sorted according to cross-sectional area or according to some other shape factor (e.g., length).

Waxing

Fruit waxing of is done to add value and improve the appearance of the fruit to the customer. Coatings also may reduce water loss and affect long-term colour changes. Waxing is achieved by passing the fruit under wax-solution sprayers, combined with rotation on expanded plastic foam rollers already soaked by the spray.

Packaging

Once sorted into grades and size, fruit are delivered by conveyor to a packing area. Here fruit may be held in rotating final size bins until they can be packed by hand into cartons. Alternatively, if automatic tray fillers are used, fruit fall into a paper pulp tray directly. Fruit and vegetables also may be wrapped individually in foam liners or films before, or instead of, packing them in a rigid container.

Figure 7-17: Pressure-pack tray system

Because it is possible to preserve food by altering their immediate environment, packaging has become an important element in preservation. Packaging fulfils several functions including containment, facilitating transportation, protection of fruit from further damage, protection of the environment from contents of package (for example, if the contents are dirty), marketing, product advertising, and stock control.

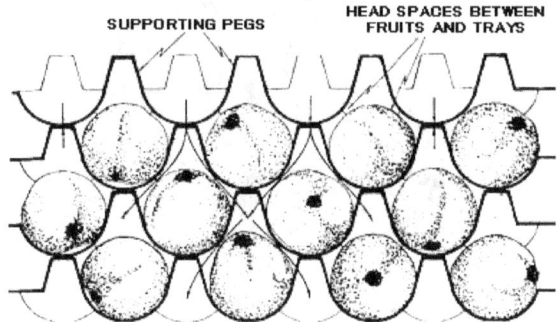

Figure 7-18: Peg-type tray system

Factors affecting packaging

Satisfactory packaging requires consideration of protection, economy, convenience and appearance. Factors affecting the choice of packaging materials include:

1. Product properties
2. Storage condition
3. Properties of economically available material.

Packaging is affected by the tendency of food to gain or lose moisture, its free or fat content, its particle size, tendency to sift and its susceptibility to spoilage by light, oxygen and organisms. Internal and external failures of both rigid and flexible containers frequently reduce quality or spoilage life of products, resistance to moisture, corrosion, leakage and package fatigue are needed to withstand high temperature or humidity or the corrosive action of certain products high in salt, fat natural acid sulphur compounds.

Figure 7-19: Sample of RPC with apples

Packages should be designed to allow cool air to flow directly over products, ensuring rapid temperature pull-down and consequent temperature maintenance. Regrettably, too many packages in current use are largely impenetrable to air movement, resulting in a slow temperature decrease, influencing quality negatively.

Figure 7-20: Sample of DRC with tomatoes

Minimizing temperature variation from the optimum recommended during storage improves uniformity in product quality. Refinements are being made in refrigeration control systems, cool-store design, and pallet-stacking patterns to optimize air temperature and airflow through and around pallets in cool stores. This prevents hot spots from developing in localized areas within the store that would lead to higher respiration rates, more rapid deterioration, and hence poorer quality than in parts of the store in which product temperature was optimized.

Packing design and packaging can also be designed to minimize water loss. To minimize condensation inside the bag and reduce the risk of microbial growth, the bags may be vented; micro perforated, or made of material permeable to water vapour. Barriers to water loss may also function as barriers to cooling, and packing systems should be carefully selected for the specific application with this in mind.

Figure 7-21: Sample of combination plastic/paper container for flowers

Packaging materials, storage or transport containers, or bins that contain synthetic fungicides, preservatives, or fumigants (or any bag or container that has previously been in contact with any prohibited substance) are not allowed for organic postharvest handling. In small-scale handling, the reuse of corrugated containers from conventional produce is strongly discouraged by organic certifying organizations.

Labeling

On the international market, some fresh fruits require considerable detail. This may include, for example, harvest date, packing date, and name of grower and packer, as well as details about the cultivar and grade. This information can be critical if quality management problems arise in the retail market.

Quality control

Quality-control checks must be carried out before the fruit leaves the packing shed. Although a check should be made once the fruit has been packed, good quality management requires that the process be continuous and cover all parts of the chain. Thus problems can be identified early and corrective action taken.

Cooling

Previously cooled fruit will have warmed up during sorting and will need recooling. For rapid cooling it is necessary to place fruit into a precooler for fast cooling to store temperature, before transferring to the normal cool store. Precoolers are designed to extract heat as quickly as possible and desirable by using forced air flow through the load to be cooled.

Types of cooling

a. *Pre-cooling:* Pre-cooling is the first step in good temperature management. The *field heat* of a freshly harvested crop — heat the product holds from the sun and ambient temperature — is usually high, and should be removed as quickly as possible before shipping, processing, or storage. Refrigerated trucks are not designed to cool fresh commodities but only maintain the temperature of pre-cooled produce. Appropriate pre-cooling methods are the most commonly used

b. *Room cooling:* Produce is placed in an insulated room equipped with refrigeration units. This method can be used with most commodities, but is slow compared with other options. Containers should be stacked so that cold air can move around them, and constructed so that it can move through them.

Figure 7-22: Refrigerated room cooling

c. *Forced-air cooling:* Fans are used in conjunction with a cooling room to pull cool air through packages of produce. Although the cooling rate depends on the air temperature and the rate of air flow, this method is usually 75–90% faster than room cooling.

d. *Hydro-cooling:* Dumping produce into coldwater, or running cold water over produce, is an efficient way to remove heat, and can serve as a means of cleaning at the same time. In addition, hydro-cooling reduces water loss and wilting. Hydro-cooling is not appropriate for berries, potatoes to be stored, sweet potatoes, bulb onions, garlic, or other commodities that cannot tolerate wetting.

e. *Top or liquid icing:* Icing is particularly effective on dense products and palletized packages that are difficult to cool with forced air. In top icing, crushed ice is added to the container over the top of the produce by hand or machine. For liquid icing, slurry of water and ice is injected into produce packages through vents or handholds without removing the packages from pallets and opening their tops.

f. *Vacuum cooling:* Produce is enclosed in a chamber in which a vacuum is created. As the vacuum pressure increases, water within the plant evaporates and removes heat from the tissues. This system works best for leafy crops, such as lettuce, which have a high surface-to-volume ratio.

Cool storage

Once at store temperature, fruit and vegetables can be transferred to a larger cool store. Its function is to maintain rather than reduce the temperature of the product.

Chilling injury

Most tropical and subtropical products are susceptible to chilling injury when exposed to temperatures above freezing but below a critical threshold temperature for each particular product. These chilling temperatures cause break down of cellular membranes, resulting in loss of compartmentalization within the cells of the tissue, increased leakiness, water soaking of tissue, and eventually pitting or browning.

Some chilled fruit fail to ripen normally, while in others there is an accelerated rate of senescence and a shortened shelf life. Symptoms of chilling injury are varied and depend on the product but include surface pitting, surface browning, internal browning to vascular tissue or in parenchyma cells, water soaking, and meatiness or wooliness of texture.

Loading and transportation

Finally, product is loaded onto appropriate transportation by forklift truck and transported to the eventual market destination. This may be days or months after packing. Although the product usually has been packaged, it is still vulnerable to damage if subjected to rough ride over a considerable distance. Over long distances refrigerated vehicles are essential.

7.7 Postharvest storage of fruits and vegetables

Fresh vegetables (including fruits) are living organisms and there is a continuation of life process in the vegetables after harvest. Changes that occur in the harvested, unprocessed vegetable include water loss, conversion of starch to sugar, and vice versa, increase flavour changes, colour changes, toughening, vitamin gain or loss, sprouting, rooting, softening and decay.

While some changes results in quality deterioration, others improves quality in those vegetables that complete ripening after harvest. To maintain the fresh vegetable in the living state, it is usually necessary to slow down the life processes, though avoiding death of the tissues, which may produces gross deterioration and drastic difference in flavour, texture, and appearance.

Vegetables for storage must be free from mechanical, insect, and disease injuries and should also be at the proper stage of maturity.

Postharvest storage system

Fresh vegetables and fruits are living organisms and there is a continuation of life processes in them after harvest. To maintain the fresh vegetable in the living state, it is usually necessary to slow the life processes, avoiding death of the tissues, which produces gross deterioration and drastic difference in flavour, texture and appearance. Methods generally employed for fruit and vegetable storage include; temporary storage, common (unrefrigerated) storage and cold (refrigerated) storage.

Temporary storage

Temporary storage, suitable for very brief storage periods, is frequently practiced in the shipping season when large lots are accumulated for car load or truck quantities. The refrigerator car or truck is a means of temporary storage while produce is in transit, short-term storage may last for four to six weeks.

Unrefrigerated storage system

Common storage system, lacking precise control of temperature and humility includes the use of insulated storage houses, outdoor cellars or mounds.

Figure 7-23: A large cellar at the Oxon Hill Manor farm in Maryland

Refrigerated storage system

Cold storage allows precise regulation of temperature and humidity and maintenance of constant conditions by use of registration. So, it can be said that cold storage is a typical storage system for fruits, vegetables, dairy products, fish, meat and other foods using refrigerated air. Cold storage allowed regulation of temperature and humidity and maintenance of constant conditions by use of a refrigeration and ventilation system.

Refrigerated storage retards the following elements of deterioration in perishable crops:

a. Aging due to ripening, softening, and textural and color changes;
b. Undesirable metabolic changes and respiratory heat production;
c. Moisture loss and the wilting that result;
d. Spoilage due to invasion by bacteria, fungi, and yeasts;

e. Undesirable growth, such as sprouting of potatoes.

One of the most important functions of refrigeration is to control the crop's respiration rate. Respiration generates heat as sugars, fats, and proteins in the cells of the crop are oxidized. The loss of these stored food reserves through respiration means decreased food value, loss of flavour, loss of salable weight, and more rapid deterioration. The respiration rate of a product strongly determines its transit and postharvest life. The higher the storage temperature, the higher the respiration rate will be.

For refrigeration to be effective in postponing deterioration, it is important that the temperature in cold storage rooms be kept as constant as possible. Exposure to alternating cold and warm temperatures may result in moisture accumulation on the surface of produce (sweating), which may hasten decay.

Figure 7-24: Refrigerated banana storage exhibition

Storage rooms should be well insulated and adequately refrigerated, and should allow for air circulation to prevent temperature variation. Be sure that thermometers, thermostats, and manual temperature controls are of high quality, and check them periodically for accuracy.

On-farm cooling facilities are a valuable asset for any produce operation. A grower who can cool and store produce has greater market flexibility because the need to market immediately after harvest is eliminated. The challenge, especially for small-scale producers, is the set-up cost. Innovative farmers and researchers have created a number of designs for low-cost structures. Optimized refrigeration and controlled-atmosphere storage conditions reduce respiration, achieving slower deterioration and enhanced storage life. Preventing ethylene contamination of the storage environment from external sources or by ethylene producing crops is important to maximize storage life. These approaches, combined with new engineering and molecular approaches will be important in the production and handling systems used to get products to market.

Factors affecting fruit storage

Under appropriate conditions, fruits can be stored for many years. It is important to establish the condition and storage period that afford the optimum balance between the cost of storage and the changes in quality of stored products.

Wide range of natural temperature and humility conditions exists in warehouses, dugouts, shelters and natural caves used for food storage. These can be highly satisfactory if conditions are sufficiently constant to allow prediction of storage life of the foods.

The temperature maintained in the above ground natural storage houses vary with site and geographical location, from below freezing to above 100 of (38°C) at relative humility from almost 100-20% or below. Temperature in caves or deeper excavations may fluctuate at less than 10°F (6 °C) annually. Utilization of these or similar sources of natural cool storage can yield advantages over above-ground storage in more uniform quality and increased life, or can reduce the cost of holding under refrigerated storage.

Postharvest storage considerations

Moisture loss prevention: Temperature is the single most important factor in maintaining quality after harvest. Temperature remains a primary concern in the storage of fruits and vegetables. The relative humidity of the storage is an important unit which directly influences water loss in produce. Water loss can severely degrade quality – for instance, wilted greens may require excessive trimming, and grapes may shatter loose from clusters if their stems dry out.

Sanitation: Sanitation is of great concern to produce handlers, not only to protect produce against postharvest diseases, but also to protect consumers from food borne illnesses. Use of disinfectant in wash water can help to prevent both postharvest diseases and food borne illnesses.

Packaging; Packaging should be designed to prevent physical damage to produce, and be easy to handle.

Effect of storage and packaging on food

Colour stability varies widely among different products and storage temperature. The only significant change at (0 -20 °C) (-18 °C-29 °C) is a slightly lighter appearance, and there is no significant change in any product at 32 °F (0 °C). Temperature of 47 °F f (8 °C) results in significant changes, particularly in the stable items. At temperature from (70-100 °F) (21-30 °C) products tend to darkened, brown, fade and lose flavour or become stale. Extremes of temperature have adverse effects on texture. Freezing causes sufficient damage to such items as tomatoes and beans, thus reduce them to substandard grades.

7.8 Fruit and vegetable processing technologies

A wide range of technology is applied to the processing of fruit and vegetables. Various processes (sometimes integrated, sometimes not) yielded one type of fruit or vegetable products Thus mangoes can be canned, dried, frozen, or made into puree, sauce, chutney, pickles, concentrate, jam, beverages etc. Papaya provides jam, pickles, chutney, nectar, toffee etc. Coconut can be processed to produce oil, milk, cream, desiccated coconut, pie fillers, soft cheese, meal, jam, soap etc. Because of the large number of products, the more common processes and those processes that pertain to some of the more significant products are highlighted.

Blanching

Blanching is a process that involves dipping the vegetables or fruits into boiling water or suspending them in steam for 1.5 to 3 minutes. This process can achieve a number of objectives.

1. Firstly, it kills microorganisms on the surface and in the case of green vegetables destroys the catalyst enzyme and inhibits the peroxidase enzyme, both of which can produce off odors and flavors in storage.
2. Secondly, it removes air from intercellular spaces (assisting in the formation of a head-space vacuum in cans) and softens tissues to facilitate filling containers.

Blanching may or may not help to preserve the colour in green vegetables. Blanching also is used to crack and loosen the skins of tomatoes, sweet potatoes, and beets prior to peeling. Fruit is not usually heat blanched because of the damage from the heat and the associated sogginess and juice loss after thawing. Instead chemicals are commonly used without heat to inactivate the oxidative enzymes or to act as antioxidants.

Canning

Canning is a method of preserving food in which the food contents are processed and sealed in an airtight container. Canning process involves cutting, washing, blanching, and then hot-loading into the can (which is usually made from tin-plated steel), which is then sealed. The can is then held at a prescribed temperature for a given time to ensure that all microorganisms are killed, before being quickly cooled. After the can is sealed, it is heat-treated, usually by immersion in boiling water or in a pressurized steam retort.

Figure 7-25: Canned products

Canning is extremely valuable, because it prevents the reentry of microorganisms completely; the life of the product (even at ambient temperature) can be several years, provided the integrity of the seal is maintained.

Freezing

Freezing is a very popular and effective means of preserving many types of vegetable. Prior to freezing, vegetables are normally blanched. Freezing is generally not very effective for cut fruit because the fruit is susceptible to browning and oxidation upon exposure to air. However, some whole fruits such as berries are frozen satisfactorily.

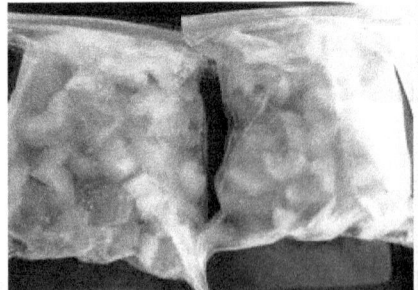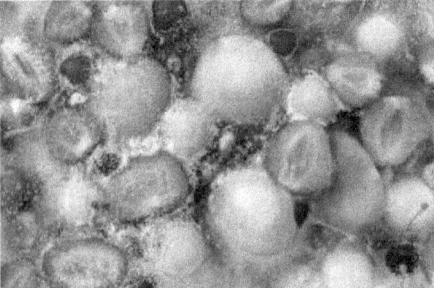

Figure 7-26: Frozen fruits

If fruits are to be of highest quality after they are frozen, they must be of highest quality in their fresh state. Freezing does not add anything to the original qualities of fresh fruits. The maturity of fruits to be frozen is extremely important in determining the quality of the product. They should be picked at the stage when they are best for eating. Choose fresh firm-ripe fruits. Do not use hard or over-soft fruit. Freeze them before they lose their freshness. If fruits cannot be frozen immediately, refrigerate them.

Freeze-drying

In freeze-drying, the product is first frozen and then placed in a vacuum-tight enclosure and dehydrated under vacuum with careful application of heat, the pressure being kept substantially below 4.6 mm of Hg.

Dehydrated   Freeze-Dried

Figure 7-27: Dehydrated and freeze dried products

The process has a number of advantages which include

1. No reduction in volume,
2. Much of the number of aqueous volatile constituents, flavours etc are retained.

Freeze-dried foods also rehydrate more rapidly than other dried foods. Although many vegetables can be freeze-dried, one disadvantage is that freeze-dried vegetables are more susceptible to oxidative deterioration than air-dried vegetables.

Drying and sulphuring

Drying is one of the oldest food processing and preservation techniques known to man. The essential feature of the process is to reduce the moisture content to a point at which enzymatic or microbial damage no longer occurs. Fruits (e.g., apricots, peaches) are halved and pitted and then laid out by hand on wooden or plastic trays.

Figure 7-28: Drying fruit products

Before they are placed in the sun to dry, or put into a high temperature drier, many fruits are first treated with sulphur dioxide, a process known as *sulphuring*. The SO_2 taken up by the fruit displaces the air from the tissue, softens cell walls so that drying occurs more easily, destroys enzymes that might darken the color, and actually produces some color enhancement. To maintain maximum quality, dried fruits should preferably be stored at 10 to 15 °C.

Desiccation

Desiccation is the state of extreme dryness, or the process of extreme drying. A desiccant is a hygroscopic (attracts and holds water) substance that induces or sustains such a state in its local vicinity in a moderately sealed container.

Figure 7-29: Desiccators

Desiccated coconut is a major product used in the confectionary industry as a bulking material. Fresh coconuts are shelled whole and the outer brown skin or testa pared off with a knife. The coconut water is drained away and the flesh is cut and washed and then sterilized in boiling water or in steam blanchers at 88 °C for 5 minutes. The pieces then are shredded into a fine wet meal in a hammer mill and dried to 25% to 30% moisture content in a steam-heated counter flow multistage drier at 77 to 82 °C for 40 to 45 minutes. Finally, the product is size-graded and packed.

Controlled ripening

A number of climacteric fruits, notably bananas, mangoes, papaya, pears, tomatoes, and avocados are picked relatively green and are subsequently ripened by introducing ethylene or acetylene gas, calcium carbide, or smoke. The fruit to be ripened ideally is placed in an airtight ripening room maintained at a constant temperature (18–21 °C for most fruits, but 29–31 °C in the case of mango). In modern practice, the fruit are packed in vented cartons stacked on pallets, and fruit temperature is controlled by forced air circulation as in a cooling facility.

Figure 7-30: Control ripening of banana

More traditional rural methods of ripening bananas and mangoes involve putting the fruit into a pit in the ground or into a heap on the surface and covering with a tarpaulin or branches. In some cases ripening is induced by placing small sachets of calcium carbide among the fruit, which in the humid atmosphere produces acetylene. Another frequently used technique is to expose the fruit to smoke produced by burning leaves or wood. Ethylene thereby is produced as one of the products of incomplete combustion. However, product quality may be reduced by this procedure, especially if immature fruit is used.

Figure 7-31: Ripening room

Controlled degreening

Degreening simply means to get rid of the green colour. As much as this procedure is simple in basis and principles, it is complicated in its application. Controlled degreening sometimes is carried out on citrus grown in the tropics. Many citrus cultivars are mature before the green colour disappears from the peel. The breakdown of chlorophyll and production of a rich orange color requires exposure to low temperature during maturation, and this explains why mature citrus frequently is sold green on markets in the humid tropics, where even night temperatures may not drop much below 25 °C.

Degreening is carried out in degreening rooms. Degreening rooms are the equipment used for the process. These rooms are totally insulated from the outside world. Green oranges are stacked in these rooms. Parameters of the optimum weather conditions for turning the colour from green to orange are then regulated, kept, and monitored around the clock. The main parameters are: Temperature, humidity, oxygen level, carbon dioxide level (CO_2) and ethylene level

Figure 7-32: Degreening process

The most rapid degreening occurs at temperatures of 25 to 30 °C but the best colour occurs at 15 to 25 °C. Fruit degreening room is mainly applied to postharvest processing for citrus, navel orange, grapefruit, lemon and other fruits. Although degreening room is not standardized as ripening room and control procedures are different, the same key environmental parameters are controlled.

Processing into purees, pastes, and edible leathers

Fruits such as tomato and mango are first peeled, destoned, and sliced, and the slices pulped in a homogenizer. The pulp then is sterilized by raising the temperature to at least 75 °C (mangoes). The puree is then canned or sealed in polythene pouches for long-term storage and marketing. Fruit "leathers" can be made by sun drying or forced-air dehydration of a puree layer on a thin plastic film.

Figure 7-33: Fruit purees and pastes

Processing into flour

A number of vegetable and fruit crops are used to produce flour, two of the major ones being soybean and coconut. Beans are first dried to 13% moisture content, cleaned, and then tempered by storing at 10% to 12% moisture content and 25 °C for 7 to 10 days and then dehulled. The dehulled beans then are conditioned by raising the temperature to 70 °C, thereby increasing the amount of oil that can be extracted. The conditioned beans then are flaked by passing them through rolls. Oil then is extracted by the solvent process. The flakes then are dried to 10% moisture content, toasted, and ground to a meal.

Aseptic processing and packaging

In contrast to canning, where the hot-filling and post sealing heat treatments are used to produce commercial sterility, in aseptic processing both the product and the packaging are made commercially sterile before the filling and sealing operations, and therefore no post sealing heating is necessary. The objective is a product that is stable for 2 to 3 months, and preferably for 6 months, without refrigeration.

Prior to packaging, the liquid or semi liquid product is heated quickly to a temperature at which it is commercially sterile and then cooled. The packages (e.g., coated cartons) are subjected to a combination of chemical and heat treatment. Higher temperatures are more effective in killing off relatively heat-resistant bacteria, while the shorter exposure time results in reduced loss of flavor, color, and nutrients. This is sometimes known as high-temperature/short-time processing.

Fermentation into alcoholic beverages, vinegar, sauces and other products

Fermentation occurs naturally in the presence of bacteria, yeasts, and molds. Bacteria produce acids, which tend to act as preservatives; yeasts produce alcohol, which is also a preservative. Fruits are fermented to produce alcoholic beverages, including apples (to produce cider), pears (to produce perry), bananas and a wide variety of berry fruits (to produce wines and liqueurs). Traditional alcoholic-beverage preparation relies on the naturally occurring yeasts and sugars to initiate the fermentation process.

Processing into jams and pickles

Four naturally occurring edible preservatives in very common use are sugar, vinegar, salt, and vegetable oil. Jams are made by boiling fruit in sugar syrup until the liquid becomes relatively stiff. Sometimes additional pectin is added to cause the jam to set to a stiffer consistency upon cooling. The jam is poured into clean, sterilized containers while hot, sealed airtight, and then cooled. In some cases it also is desirable to cover the surface of the jam with a thin layer of wax. Pickling of vegetables and unripe fruit (such as papaya and mango) usually involves some heating with the addition of spices, sugar, and other flavorings, followed by immersion in vinegar (acetic acid) or oil in glass jars.

Figure 7-34: Jams

Irradiation

Irradiation is the process by which an object is exposed to radiation. The exposure can originate from various sources, including natural sources. Small doses of ionizing radiation (electron beam processing, X-rays and gamma rays) may be used to kill bacteria in food or other organic material, including blood. Ionizing radiation treatment has four potential uses, namely the disinfestations of insect pests, extension of shelf life by retarding ripening and sprouting, inhibition of rotting, and disinfection of material affected by harmful organisms such as salmonella.

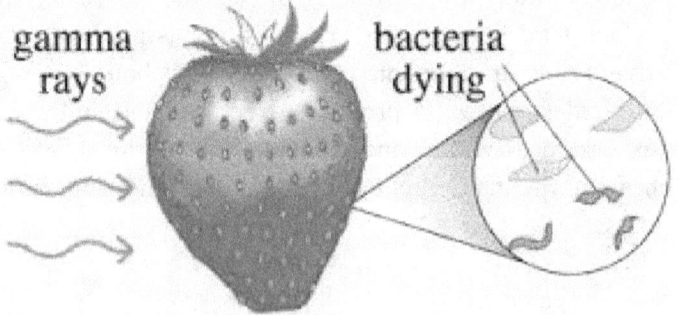

Figure 7-35: Ionizing radiation process

Irradiation currently is used mainly for inhibition of sprouting in potatoes and onions. At present Japan makes the widest use of irradiation for these products. Health concerns, in particular the relatively unknown effects of free radicals and other chemical changes produced by the irradiation process, will continue to put limits on the application of irradiation to foods.

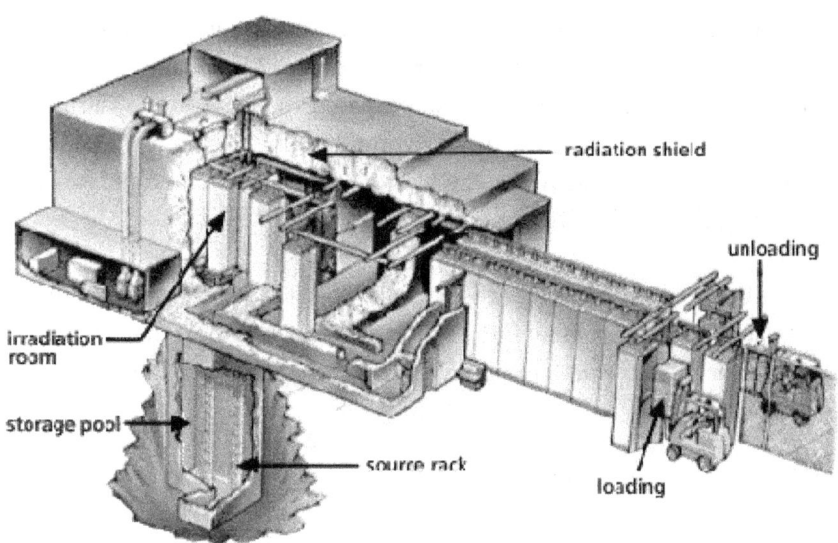

Figure 7-36: Irradiation process plant

7.9 Recent developments in fruit and vegetable processing

The number of varieties of fruit products available to consumers has increased substantially in recent years. The fruit industry has benefited from the increased recognition and emphasis on the importance of these products in a healthy diet. Traditional processing and preservation technologies such as freezing and drying together with the more recent commercial introduction of chilling continue to provide the consumer with increased choice.

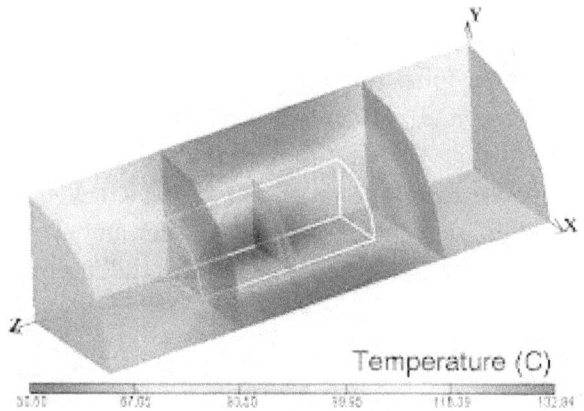

Figure 7-37: Microwave ohmic heating

New heating methods are being developed for example, Ultra heating technology (UHT), microwave ohmic heating, and freezing methods such as cryogenic technologies combined with new packaging materials and technologies e.g. aseptic packaging, modified atmosphere packaging etc The overall trend in new fruit products is "added value", thus providing increased convenience to the consumers by having much greater varieties of ready prepared fruit products. Fruit juices in confectionery products are now left up to the imagination of the manufacturer. These products must of course hold up to the standards of flavour integrity and product excellence during the shelf life of these products.

Each new fruit processing center needs a good, specific preliminary study including among others raw material availability, harvesting and transport means, marketability of finished product and training at all levels in processing of indigenous crops which must be expanded.

New postharvest opportunities and advances

Customer satisfaction

Customers, especially supermarket buyers, are requiring increasingly uniform and consistent products with narrowing margins of quality specification. Technologists are poised to make further significant contributions to improving the quality of fresh harvested horticultural products. Further refinements can be expected as we learn more about the interaction of pre harvest factors with responses of products to the postharvest environment.

Postharvest treatment

Specific knowledge of product response to carbon dioxide (CO_2), Oxygen (O_2), and ethylene (C_2H_4) at different times and temperatures after harvest, together with the sophisticated temperature- and gas-control systems currently available, creates the opportunity for modulating storage environments to continuously minimize respiration rate and hence deterioration. This can be achieved in both static land-based cool stores and in sea going containers. Intermittent shock treatments to minimize some physiological disorders, including chilling injury, also may be possible.

Improved packaging system

Considerable efforts are being made to develop active packaging to overcome problems associated with changing temperatures during storage and transit of modified atmosphere packages. Development of films that can increase permeability in response to a chemical signal (e.g., ethanol or acetaldehyde produced during anaerobic respiration) may soon become a reality. The possibility for creating surface coatings, which reduce water loss but at the same time allow the establishment of appropriate internal atmospheres in fruit but do not allow physiologically "dangerous" levels of gases to develop, is exciting.

Equipment sophistication

Rapid advances are being made in creating equipment that can monitor internal quality attributes nondestructively. Already it is possible to grade fruit electronically for size and colour (Israel experience); near infrared detectors already are used commercially in Japan and France to measure soluble solids content in fruit on grading lines. The ability to separate products accurately and consistently on the basis of specific internal chemical composition creates the opportunity for providing particular quality attributes (taste or sensory) for discerning niche markets.

Biotechnology

Molecular biology has allowed scientists to incorporate desirable genes into the genetic makeup of traditional crops. Genetic manipulation of plants probably offers the greatest opportunity for making long-term sustainable advances in reducing deterioration rates and extending storage and shelf life of horticultural products (Biotechnology). Naturally occurring, long-life mutant tomatoes are now available for commercial production. Combining genetic traits conferring normal shape and flavor attributes with those conferring reduced ethylene production, traditional plant breeders have been able to create highly acceptable cultivars that have a much extended shelf life compared with previous cultivars.

The first commercially produced food plant of this type accepted by the Food Drug and Administration (FDA) in the USA was the Flavr Savr tomato produced by Calgene. Molecular technology was used to inhibit cell-wall breakdown and hence slow fruit softening. The Flavr Savr tomatoes are more resistant to handling damage, and hence postharvest decay, than normal tomatoes, as well as producing a more viscous paste that is a great advantage for tomato processors. Tomatoes have been produced with much reduced ethylene production and hence greatly extended shelf life. Similar lines have been created for apple, peach, and kiwifruit and are being evaluated for other production and quality attributes.

Value addition

Postharvest technologies have brought about the emergence of value added products with improved taste, flavor, and increased shelf life, resistance to insect and pest attack. For instance National Roots Crops Research Institute (NRCRI), Umudike processed cocoyam into various value added products such as chips, soup thickeners etc.

7.10 Fruit processing equipment for unit operations

1. *Size-reduction units*

Size reduction processes are carried out with a range of cutting elements including choppers, cutter, slicer, dicers and shredders.

Slicing: Slicing is achieved when rotating or reciprocating blades cuts into the product as it passes beneath.

Dicing: In dicing operation, the material is sliced and then cut into strips by a series of rotating blades. The strips are then cut into cubes by a further set of knives acting in a perpendicular plane.

Shredding: Shredding is performed either by multiple rotating knives (in a machine similar to a hammer mill, but with the hammers are replaced by knives) or in a squirrel-cage shredder, in which the product is made to pass between two counter rotating cylindrical cages, both fitted with knives.

Colloid mills: Colloid mills are essentially disc mills, with one disc stationary and one disc rotating at 3000 to 15000 revolutions per minute. The small clearance between them (0.05–1 mm) produces high shearing rates and forces. Colloid mills are particularly suitable for high-viscosity liquids. The discs may be flat, conical, or corrugated or have other asperities on them that are suited to processing a particular material.

2. *Extraction units*

Juice extraction: Two types of juice extractors are used. The first is a mechanized and automatic version of the standard kitchen lemon press, in which the fruit is cut in half and squeezed against a ribbed hemisphere. In the second type, the whole fruit is held between a lower and an upper cup, while cutters in the center of both cups begin to cut plugs in the fruit.

Pressure then increases on the fruit, forcing the material through the bottom plug and into a prefinisher tube. The peel then is discharged between the upper cup and the cutter (Figure 7-38). The juice and juice sacks then pass through small holes in the finisher tube, while larger particles are discharged through the bottom of the tube.

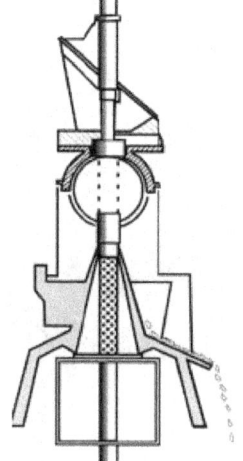

Figure 7-38: Juice extractor

Homogenizers: Homogenizers come in a number of designs. In pressure homogenizers, a high-pressure pump operating at 10 to 70MPa feeds the liquid through a poppet valve with the poppet set to produce a gap of 300 μm. The rapid drop in pressure and sudden change in velocity produces turbulence that reduces the globule size. In some equipment, the high-velocity jet is made to impinge on a breaker ring to achieve a better result. A second type of homogenizer (used for ice cream, salad creams, and essential oil emulsions) uses ultrasonics.

Figure 7-39: Homogenizers

Centrifugal clarifiers: Clarifiers are used to separate solids from liquids. The simplest form of clarifier is the cylindrical-bowl type. The liquor to be clarified (usually with a maximum of 3% w/w solids) is introduced into the bowl, and solids form a cake on the bowl wall. When the cake has built up to a given thickness, the bowl is drained and the cake removed through an opening in the bottom of the cylinder.

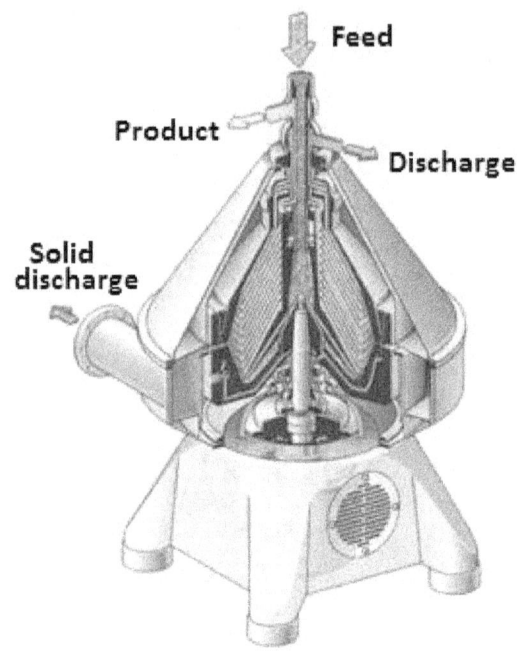

Figure 7-40: A section of a centrifugal separator

Liquors that have higher solids content are clarified using centrifuges that have bowls of a biconical shape. They are of two types, nozzle centrifuges and discharge centrifuges. The nozzle type has small holes at the periphery of the bowl, through which solids are continuously discharged. In the valve type, the holes are fitted with valves. These valves periodically open for a fraction of a second to discharge the solids that have accumulated. Figure 16-shows a section of a biconical clarifier.

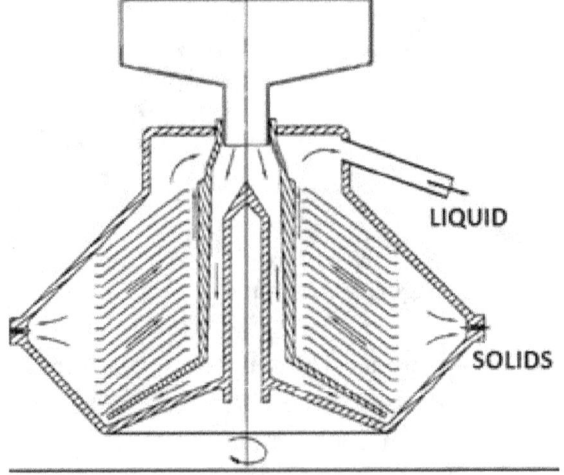

Figure 7-41: A section of a biconical clarifier

3. *Juice filters*

Plate-and-frame filter press: A very common type of filter is the plate-and-frame filter press. Cloth or paper filters are supported on vertical plates, a number of which are clamped together depending on the filter capacity required. The feed liquor is pumped into the press under pressure. Having passed through the filter cloths, it flows down a number of grooves in the surfaces of the plates and drains out through an outlet channel at the bottom.

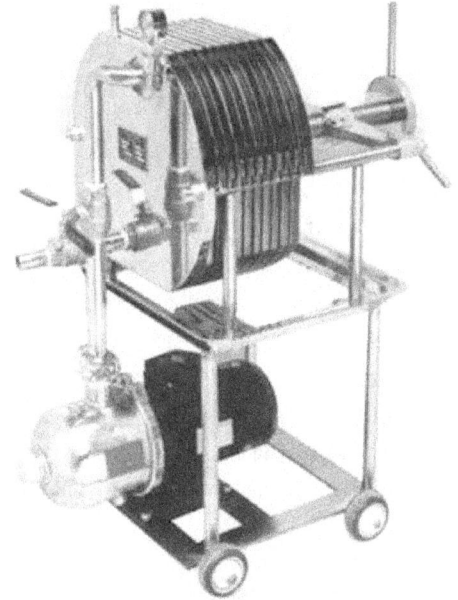

Figure 7-42: Orange juice Plate-and-frame filter press

As with all filters, as a bed of particles collects on the filter cloth, the pressure required to maintain the flow increases. Once the pressure has increased to a predetermined value, the plates are back-washed with water. To remove the cake fully, the press is dismantled, the plates cleaned, and the press reassembled. This is obviously a time-consuming operation.

Shell-and-leaf pressure filter: Another type of filter is the shell-and-leaf pressure filter. It consists of meshing leaves, which are supported on a hollow frame. The mesh is coated with a filter medium. The leaves are enclosed in a pressure vessel. Feed liquor is pumped into the pressure vessel, and solid particles collect on the filter medium. When the filter is choked with cake, the cake is blown off or washed from the leaves. Thus the cleaning is much less labour-intensive than for plate filter presses.

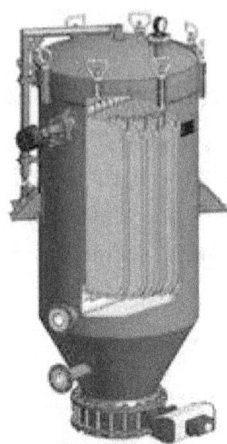

Figure 7-43: Vertical pressure leaf filter

Rotary-drum vacuum filters: Rotary-drum vacuum filters are able to work continuously. They consist of a cylindrical drum that rotates slowly about a horizontal axis and is roughly 50% submerged in a bath of liquor. The drum is divided into a number of shallow airtight compartments, each compartment being covered in filter cloth and connected to a central vacuum pump. As the drum rotates, cake builds up. When the cake layer lifts out of the liquor, it is washed and drained. When each compartment reaches a certain position, the vacuum is removed and air pressure applied to loosen the cake. Finally the cake is scraped from the filter cloth on a continuous basis.

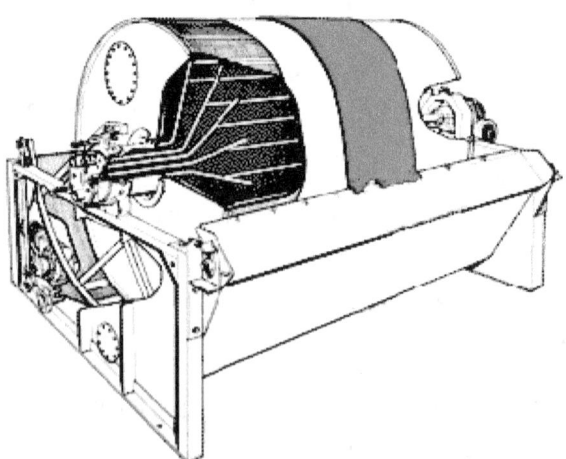

Figure 7-44: Rotary-drum vacuum filter

4. Presses

Ram press: The ram press is a batch press, frequently used for small-scale oil extraction or grape juice expression. Pulp or oil-bearing material is placed into a heavy-duty perforated metal cylinder or slatted cage and a pressure plate forced down onto the material by a screw or hydraulic mechanism. In some instances, the pulp or oil-bearing material is held in a cloth bag. Liquid flows through the perforations or slots and is collected.

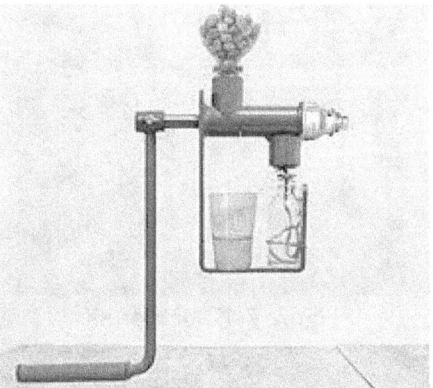

Figure 7-45: Ram press

Roller press: Roller presses allow fruit pulp to be pressed on a continuous basis. The pulp passes between two heavy fluted metal rollers. The rollers also may be hollow and perforated. Juice is expelled along the flutes or through the perforations, while the cake is removed from the rollers by scraper blades. With this type of press, high pressures can be achieved if required.

Figure 7-46: Roller presses for different applications

Belt press: A further development from the roller press is the belt press. The belt (normally plastic) passes over two stainless-steel cylinders, one perforated and the other not perforated. Fruit pulp is loaded onto the inside of the belt and is squeezed between the belt and the perforated cylinder. Juice is expelled through the perforations, while the press cake is scraped off the belt at the other end.

Figure 7-47: Belt press

Screw press (expeller): Screw presses consist of a stainless-steel helical screw inside a strong housing. The pitch of the screw decreases towards the outlet end, resulting in a gradually reduced flow area and an increasing pressure applied to the pulp. Holes in the cylinder allow the juice or oil to escape. At the outlet end, the flow is choked by a conical plug that can be adjusted to vary the pressure exerted by the screw.

Figure 7-48: Screw press

The screw mechanism causes considerable friction, resulting in heavy power consumption and heating of the material. Higher temperatures usually improve oil-expelling efficiency by reducing viscosity, and so this heating is beneficial. Additional heating sometimes is used by wrapping electrical heaters around the cylinder. Some of the larger screw presses have twin, contra rotating screws.

5. Extruders and extrusion cookers

Generally, extruders are similar to expellers in that they are of the single screw or twin-screw type, demanding high energy input and sometimes generating considerable frictional heat. At the outlet, they have a die of the required shape, through which the material is forced, and then rotating knives (or a guillotine) cut the extrusions to the right length or shape. Thus strips, rods, tubes, spheres, doughnuts, and shells can be produced. Frequently the extruded material subsequently is sprayed with sugar solution or flavouring.

Figure 7-49: Extruder and extrusion product

In cold extrusion, the extruder has a deep-flighted screw that rotates at 30 to 60 revolutions per minute. It is used for soft, doughy material that produces little heat. Special dies can be made to inject a filling into an outer shell, a process known as coextrusion.

In extrusion cookers, the food are heated within the extruder by a steam jacketed barrel or a steam-heated screw. The residence time is longer (30–90 sec). The rapid release of pressure as the material emerges from the die causes sudden expansion of steam and gas within the material, yielding a highly porous, low-density product.

6. Blancher

Blanching can be done by either a batch process or a continuous process. Figure 7-50 shows a sketch of a continuous steam blancher.

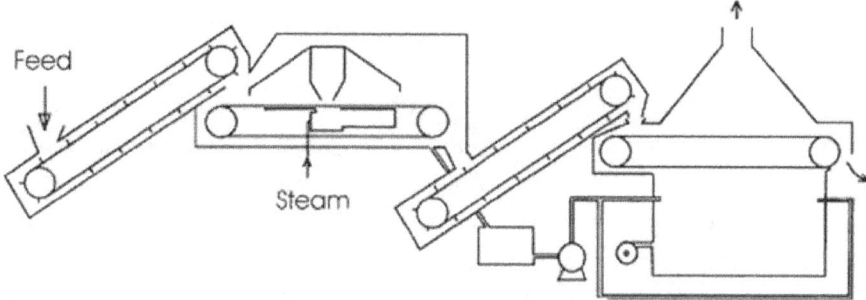

Figure 7-50: Continuous steam blancher

Steam-heated and water-heated kettles: Heating of liquids and slurries frequently is done using stainless-steel kettles, which consist of a hemispherical pan having a jacket around them through which saturated steam or hot water is passed.

7. Evaporators and dehydrators

Open and closed pan evaporators: These are essentially similar to the kettles described previously and can be steam heated or heated by gas or electric heating elements. They are cheap but relatively energy-inefficient and labor-intensive.

Short-tube evaporators: These are essentially tube-and-shell heat exchangers. The heat exchanger contains a vertical bundle of tubes inside a vessel or shell. Steam condensing on the outside of the tubes heats the feed liquor, which rises up through the tubes, boils, and recirculates through a central passage called a downcomer.

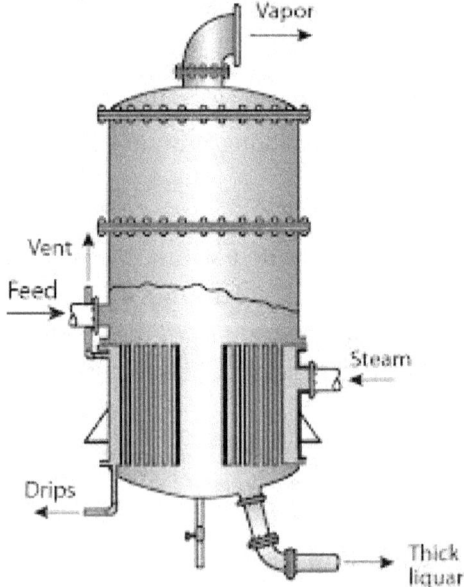

Figure 7-51: Short-tube evaporator

Long-tube evaporators: These consist of a vertical bundle of tubes encased in a steam shell which can be up to 15 m high. Liquid is introduced into the evaporator at just below boiling point. Within the tubes, boiling commences and the expansion of steam forces a thin film of liquor along the walls of the tubes, rapidly concentrating as it goes. For heat-sensitive, viscous liquids the feed is introduced at the top of the tube bundle. This type of evaporator is relatively energy-efficient, and it is widely used for citrus-fruit juices. Other evaporator types include the external calandria type, forced-circulation type, plate type, expanding flow type, and mechanical thin-film evaporators.

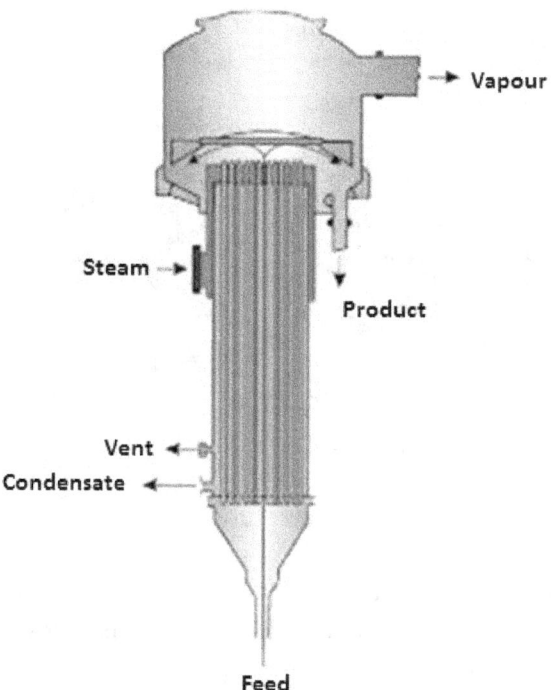

Figure 7-51: Long-tube evaporator

Retorts and sterilizers: Retorts are pressurized containers in which canned products are sterilized or heat processed. With this batch process, the cans are loaded into the retort in a special cage, and the retort sealed with a gastight lid. Saturated steam is introduced into the retort, and as it condenses onto the outside of the cans the latent heat raises the temperature of the contents rapidly. After heat treatment, cooling water is passed through the retort. Once the product temperature has dropped below 100 °C, the pressure is released and cooling continues to 40 °C. Batch-type retorts have certain advantages but are labor-intensive. Continuous-pressure sterilizers are in use with capacities well above 1000 cans per minute.

8. Plate heat exchanger

These are widely used for pasteurization of low-viscosity liquids such as fruit juices. They consist of a series of thin vertical stainless-steel plates held tightly in a rigid frame. The gaps between the plates form narrow channels, and alternate channels are connected together in parallel. The liquid to be sterilized is passed through one set of alternate channels, while the heating medium (steam or hot water) is passed in between.

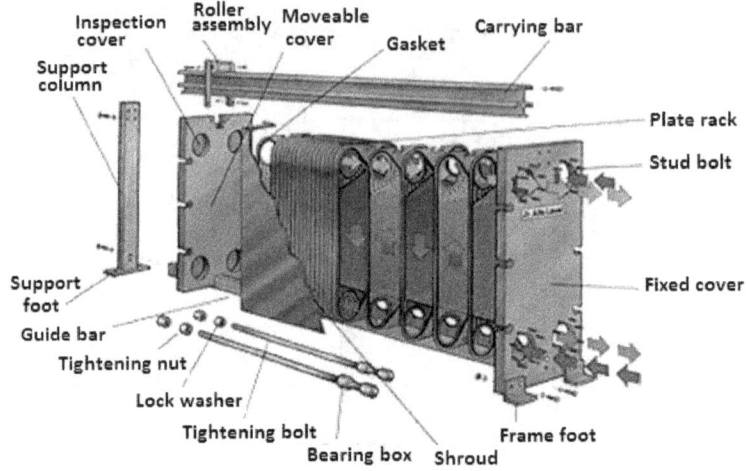

Figure 7-52: Long-tube evaporator

The inlets and outlets are arranged to form a counter flow pattern. Plates usually are corrugated to cause fluid turbulence, thereby reducing the thickness of the boundary layer and increasing the rate of heat exchange.

Further reading

ASAE Data: ASAE D2704 (1980). Design of ventilation systems for poultry and livestock shelter. Agricultural engineers yearbook 379-397.

ASHRAE. 1997. *Ashrae handbook, fundamentals, si edition*. Atlanta, GA: American Society of Heating, Refrigerating and Air-conditioning Engineers.

Bencini, M. C. 1991. Post-harvest and processing technologies of African staple foods: A technical compendium. FAO Agricultural Service Bulletin 89. Rome: FAO.

Bill Wilcke, (1999) Using Flat Buildings for Dry Grain Storage. University of Minnesota Extension Services. http://www.bae.umn.edu/extens/postharvest/

Bott, E. W., and S. Scottler. (1989). Centrifuges, decanters and processing lines for the citrus industry. Technical-Scientific Documentation No. 14,Westfalia Separator AG, Oelde, Germany.

Boumans, G. 1985. *Grain handling and storage*. New York: Elsevier.

Brooker, D. B., F.W. Bakker-Arkema, and C.W. Hall. (1992). *Drying and storage of grains and oilseeds*. New York: Van Nostrand Reinhold.

Croissant R.L. (1998) Managing Stored Grain Colorado State University Cooperative Extension. 9/92. Reviewed 9/98. www.colostate.edu/Depts/CoopExt

De Leon, S. Y. (1990). In *coconut as food,* ed. J. A. Banzon, O. N. Gonzalez, S. Y. de Leon, and P. C. Sanchez, Philippines Coconut Research and Development Foundation

Ensminger, S. (1991). *Feeds and nutrition.* San Francisco: Ensminger.

Farm structures fact sheet, (1986). Grain, forage and feed structures the Canada plan service.

Fellows P. (1988). *Food processing technology: principles and practice,* pp. 201–209. VCH Publishers.

Hotchkiss, J. H. (1989). Aseptic processing and packaging of apple juice. In *processed apple products,* ed. D. L. Downing, pp. 189–212. Van Nostrand Reinhold.Wills, R. B. H., W. B. McGlassen, D. Graham, T. H. Lee, and E. G. Hall. (1989). *Postharvest: an introduction to the physiology and handling of fruit and vegetables.* New South Wales University Press, pp. 117–118.

Ingram, J. S. 1972. Cassava processing: commercially available machinery. Tropical Products Institute Publication No. G75. London: Tropical Products Institute.Joseph P. Harner III, Timothy J. Herrman, Carl Reed, (1998).Temporary Grain Storage Considerations. Kansas State University, MF-2362

Kwatia, J. T. 1986. Rural cassava processing and utilisation centers. UNICEF-IITA Collaborative Program for Household Food Security and Nutrition Report. Ibadan, Nigeria: IITA.

Lennart P. B. and Whitaker J. H. (eds) (1998). Farm structures in tropical climates. A Textbook for Structural Engineering and Design. FAO/SIDA Cooperative Programme. Rural Structures in East and South-East Africa Food and Agriculture Organization of the United Nations

Microsoft® Encarta® Reference Library 2002. © 1993-2001 Microsoft Corporation. All rights reserved

Murray, D. R. (1990). Biology of Food Irradiation, pp. 48–53. Research Studies Press and John Wiley & Sons.

Pierce, R. O., and B. A. McKenzie. (1984). Auger performance data summary for grain. Paper No. 84-3514. St. Joseph, MI: ASAE

Reyes, V. G., L. Somons, and C. Tran. (1995). Preservation of minimally processed carrots by edible coating and acid treatment. Proceedings Australasian Postharvest Horticulture Conference, Melbourne, Australia, September 1995, pp. 451–456.

Reyes, V. G., and I. V. Gould. (1995). Improved processing and packaging of selected minimally processed vegetables. Proceedings Australasian Postharvest Horticulture Conference, Melbourne, Australia, September 1995, pp. 445–450.

Salunkhe, D. K., and S. S. Kadam. (1995). *Handbook of fruit science and technology*. Marcel Dekker.

Wilson, J. E., and L. Victor. 1980. Relationships between seedlings and their vegetative progenies in white yam (*d. Rotundata*). *Inra int. Seminar on the yam 1981*, pp. 269–278

Part 3

FARM STRUCTURAL
REQUIREMENTS

CHAPTER 8

Farmstead Layout and Planning

8. Introduction

The traditional life style of the farming communities of tropical Africa and other parts of the world is undergoing many changes. People are becoming better educated, coming into contact with other cultures and technologies, and are gradually losing their knowledge of the traditional crafts and agricultural methods that were practiced by their ancestors. This is encouraging a change from the traditional way of life to a more modern, in some respects, westernized, mode of living with a desire for appropriate farmstead.

Figure 8-1: A hunter-gatherer society

Planning the design and construction of a farmstead requires decisions with which the farm family must live for a long time, perhaps a lifetime. These decisions are likely to be highly personal because of individual preferences, financial situation, family size, location and other circumstances. There are a number of factors to be considered and questions to be answered before building a farmstead.

8.1 Farmstead & layout

A farmstead is a farm containing buildings used for commercial agriculture. The farmstead forms the nucleus of the farm operation where a wide range of farming activities takes place. It normally includes the dwelling area, animal shelters, storage structures, equipment shed, workshop and other structures.

Farmstead layout

Farmstead layout is the characteristic plan of area view of land indicating zonal distribution of farm resources such as working buildings, machinery, social amenities etc.

Zone layout of farmsteads

Layout of any typical farmstead is grouped into four zones, each 10 to 30 meters wide by concentric circles as shown in Figure 8-2. The advantage of zone planning is to provide space for present farm operations, future expansion and a good living environment. However, in many African cultures the livestock house has traditionally been placed at the centre of the farmstead. Thus the zone concept runs counter to tradition and may not be desirable.

Zone I: *Family living area*: This area is located at the centre of the farmstead; it should be protected from odour, dust, flies, etc. Recreation and infrastructure facilities such as clinics, lawns, garden, guest parking lots etc must be located close to the living quarters.

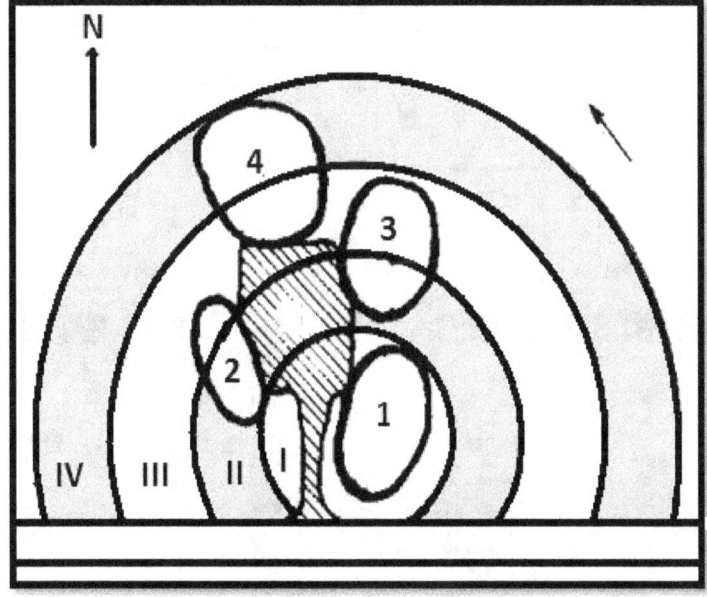

Figure 8-2: Farmstead zone layout

Zone II: *Machinery center*: This must not be too close to living quarters because of noise and dust pollution and should not be far from storage, farm production center and other related services.

Zone III: *Livestock and equipment area*: Graining, feed mill and other livestock houses are located far from farm houses because of dust, noise and odors. Grain and feed handling equipments and other processing plants requiring electric power and good voice assess are located here.

Zone IV: Major livestock facilities weather confined to a building or on dry lot demand for adequate space, drainage, waste management asses, loading facilities, feed distribution and other services. Again farm can become a livestock farm or vise visa so allow for both grain and livestock production in your master plan design to protect future growth, efficiency and sales value.

8.1.1 Patterns of farmstead layout

Figure 8-3 shows an ideal farmstead layout with shop, machinery, livestock, grain and feed storage areas indicated.

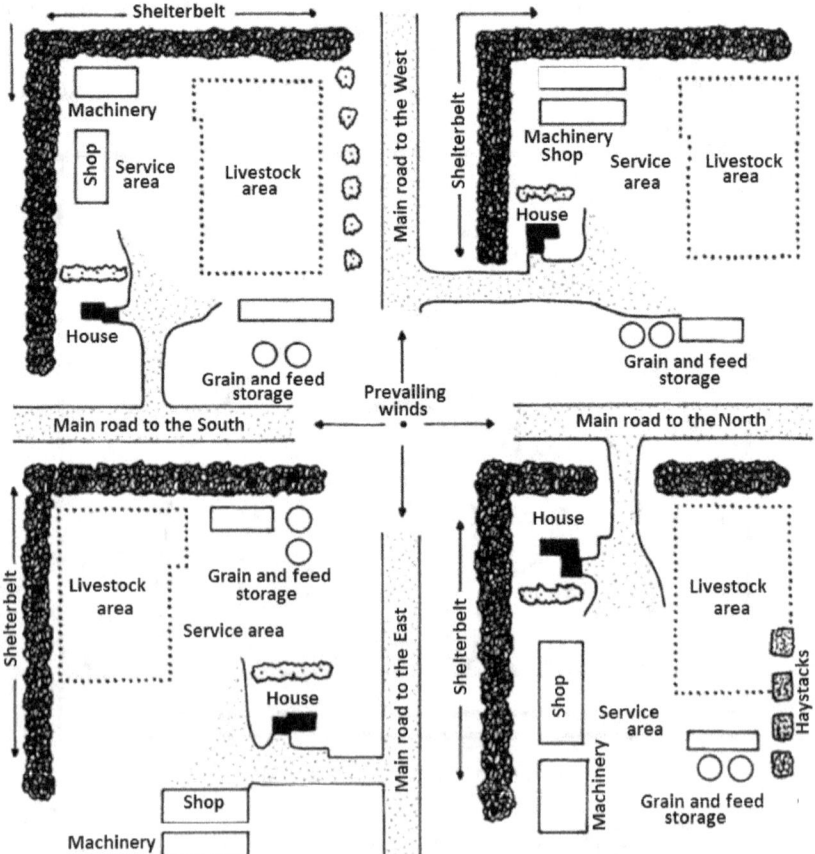

Figure 8-3: Farmstead layouts in relation to main road

Types of farmstead layout

1. Dispersed layout

A dispersed farmstead layout comprises of loose clusters of buildings with no evidence of formal planning and often no clear principal yard area.

Dispersed layout characteristics: This type of plan is characterized by open views from the surrounding landscapes into the farmsteads, because the buildings typically face different directions and sometimes overlooking two or more access points.

- Working buildings are generally ranged alongside routeways leading into the farmstead, face into their own yard areas or face into areas shared by other buildings.
- Dispersed layouts are often dissected by public rights of way which provide access into the heart of the farmstead.

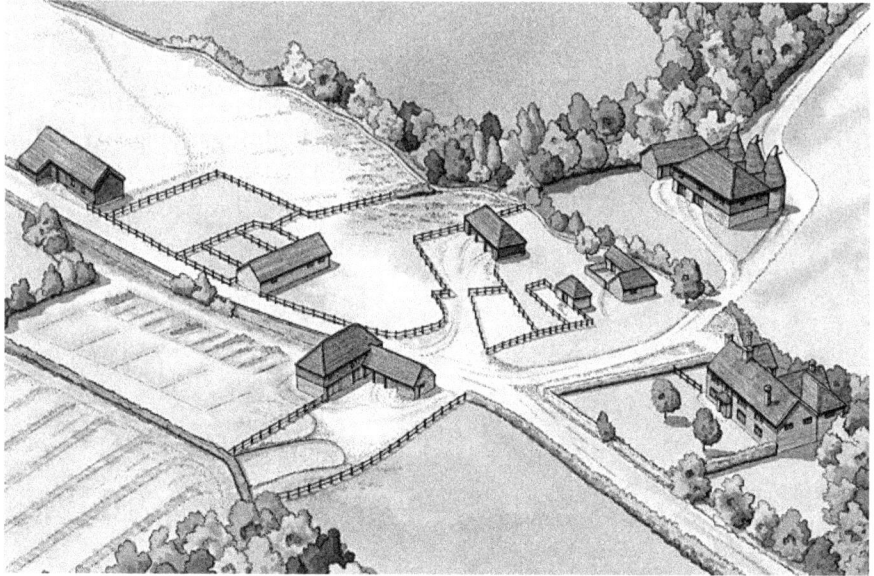

Figure 8-4: Dispersed plan layout

Dispersed plans display an enormous variation in their scale and pattern. The major types identifiable are:

- *Dispersed clusters*: Clusters of loosely-arranged buildings comprise the basic building block of farmsteads. The smallest clusters of house and barn are most strongly representative of the development of the majority of small farms up to the 18th century. They can also comprise very extensive groups of buildings, including rows of linked buildings.

- *Dispersed multi-yard layout*: These can be large in scale, and comprise of small farmsteads which are dispersed in their overall form but include two or more clusters of buildings, typically piecemeal in their development, which are arranged around and usually face into working yards. They can also contain short rows of linked buildings.

- *Regular multi-yard layout*: These can also be large in scale, and are distinguished by the regular layouts of the multiple yards.

- *Dispersed driveway layout*: These are arranged along wide driveways or tracks and may include buildings with small yard areas, short rows of linked buildings and free-standing buildings

Row type layout

Row layout comprises of long ranges of buildings, often with a series of separate yards. Some larger examples consist of two rows of buildings lying parallel to each other.

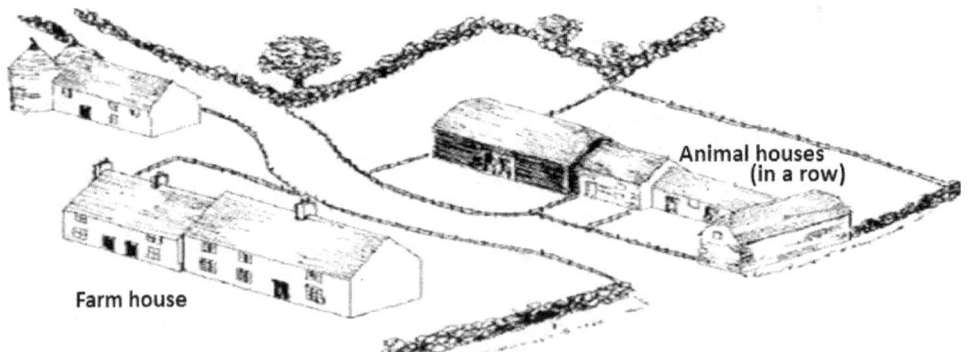

Figure 8-5: Row type plan layout

Layout characteristics

This type of layout is very rare and provides an example of a farmstead type that responds to the routes and tracks that define the character of the landscape. Some of its characteristics include:

- Building placed to one side or within the main group.
- Animal houses can be faced towards or away from main routes and tracks.

Regular layout

This comprise of farmsteads dominated by a regular arrangement of linked buildings, often of a single build, around yards. Regular courtyard elements may also be found as elements of dispersed multi-yard and driftway plans.

Layout characteristics

- Regular layouts are not a major characteristic of farmsteads but generally dated from the late 18th and 19th centuries and are strongly concentrated in landscapes enclosed and re-planned in the late 18th/19th centuries.
- Regular layout plans are more likely than other farmstead groups to have a multi-functional range of buildings for storage and processing rather than a barn.

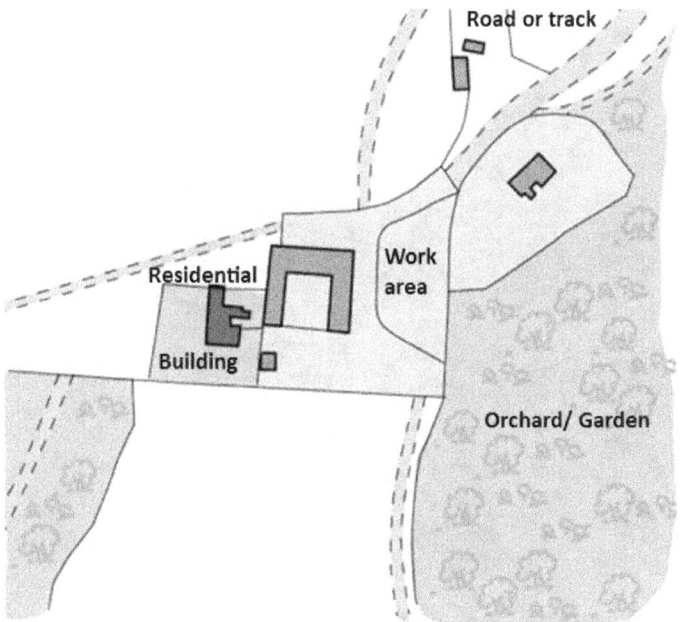

Figure 8-6: Regular U-shaped layout plan

- Some of the regular plans (particularly L-plans) may be derived from loose courtyards where a cattle shelter has been added to an earlier barn. Therefore, the number of true regular courtyard L-plans is likely to be lower than the figures suggest.
- The lowest proportion of pre-1750 farmstead buildings found on these farmstead types. Examples of whole farmsteads with regular plans which include older buildings, and thus result from piecemeal development rather than wholesale rebuilding, are rare.

- Regular courtyard farmsteads often display greater consistency in the use of materials and constructional detail. They are also more likely to use non-local materials such as Welsh slate for roofing.

Linear, parallel and attached-L plan layouts

These types layout are rare and represent less than 1% of recorded farmsteads. These are the plan types most difficult to identify from historic mapping, as they are typically associated with small farmsteads. They can include early surviving buildings or may be later farmsteads developed in areas of poorer soils such as heartland areas. Most farmsteads of this plan type have been removed from agriculture.

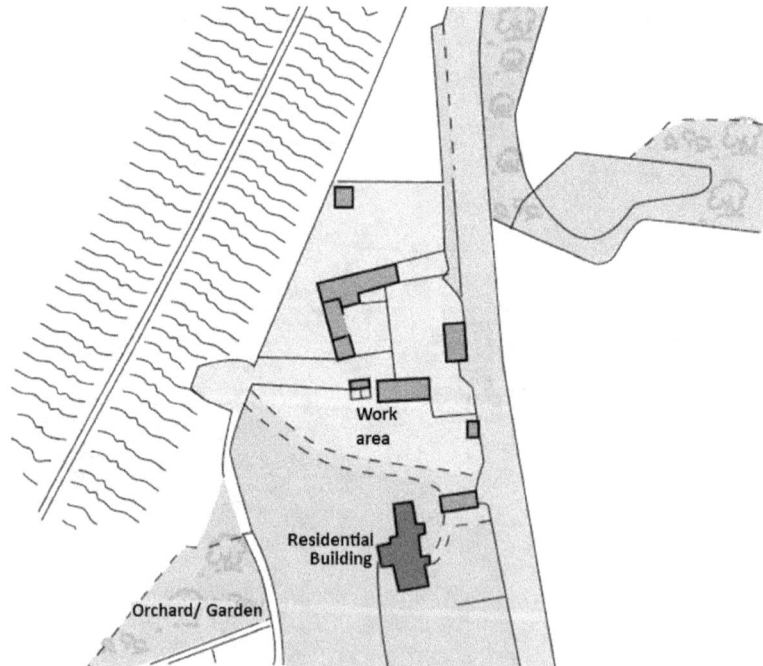

Figure 8-7: L-plan layout

Layout characteristics

- Linear plans have the farmhouse and a farm building, usually a barn, attached in-line.
- Attached L-plans have the house and working buildings attached to each other in an overall L-plan.
- Parallel plans have farmhouse and an agricultural building lying parallel to each other with a small yard area between.

- Parallel plans in the high Weald typically have the agricultural building behind the farmhouse.
- Linear and attached L-plans with unconverted agriculture buildings are very rare.

Loose courtyard layout plan

Loose courtyard layouts are a major characteristic of high farmsteads, representing 45% of mapped farmsteads. They comprise farmsteads whose working buildings - primarily the barn and cattle housing - are focused on a cattle yard, with buildings on one or more sides. Some of these plans may have developed from earlier dispersed cluster plans.

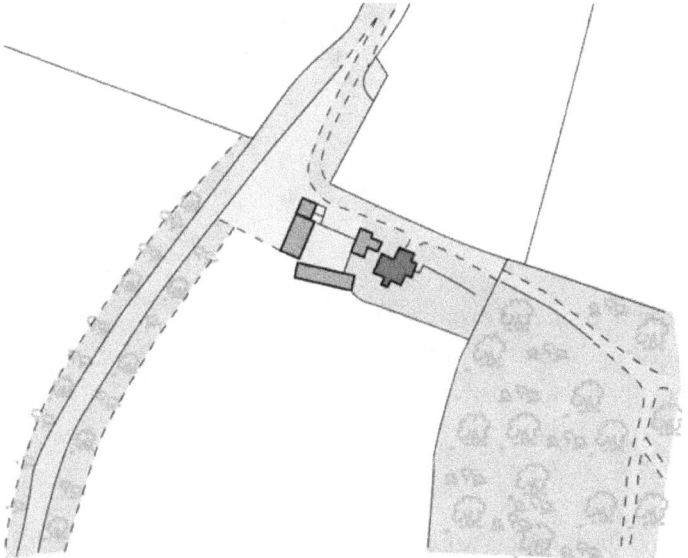

Figure 8-8: Loose courtyard plan

Layout characteristics

- Timber-framed barns are the dominant building of loose courtyard plans and are usually accompanied by cow sheds and/or a stable. Loose courtyard plans with two barns are rare.
- Buildings typically face into cattle yards, the external elevations having few if any openings.
- Public rights of way often pass close by but rarely cut through loose courtyard farmsteads allowing a greater degree of privacy and coherence than many dispersed plan farmsteads.
- Smaller and ancillary buildings set away from the yard are common - particularly coast houses, stables and cartsheds facing towards routes and tracks.

8.2   Farmstead planning

Planning: Planning is a very complicated method and procedure of trying to approximate the time and cost of a series of activities (often referred to as *'estimation'*), as well as materials using a set of pre-defined resources. Planning is an important activity which determines the success of any endeavour. It is often said that "without plans, purposes are frustrated". Planning is the first and most important step in designing a farm stead.

Farmstead planning: Farmstead planning is the mapping out and distinguishing of needed facilities, allowing as much flexibilities as possible for future changes on the farm. It involves seeking financial advice on making a cost (returns) analysis and evaluating the feasibility of project work plan which may require the services of a major programmer.

Objectives of farmstead planning

These includes,

1.   Provision of room for expansion,
2.   Improved performance, and
3.   Higher capacity and better use of labour.

Planning a farmstead

In planning four critical factors are required in any successful farmstead development:

1.   *Identification of goal*. The purposes of setting up a farmstead should be well lay out and understood.
2.   *Site selection*: Choice of farmstead location is dependent on some factors among which are:

   a.   *Drainage*: Well drained soil is necessary for some operations and also for the purposes of efficient waste disposal and setting up of waste disposal structures such as septic tank.
   b.   *Waste management*: - The ability to handle waste without problems is very important. This is particularly so if the farmstead will house a major livestock enterprise. The site must conform to all state and local environmental regulations; the topography must be satisfactory for the required storage and

drainage of manure and other effluents; prevailing wind direction is required to prevent pollution or dust from mills etc.

c. *Water*: Availability of good quality of water for the farm is very important, almost all activities on the farm require water and it must be available in adequate quantity.

d. *Utilities and services*: These include telephone, electrical services, school bus, product delivery and pick up, access drives etc. The soil should be well drained and rich enough to provide landscaping gardens, play areas etc.

e. *Building orientation*: - Air drainage and maximum sunshine may require orientation on a gentle southerly slope. Prevailing winds must be considered and natural barriers used where possible.

f. *Expansion opportunities*: Adequate provision for future expansion must be provided for. Growth in this farm stead enterprises should be anticipated and the layout should facilitate expansion of buildings and services. It is pertinent to also provide for expansion of all facilities such as machineries, utilities etc. It is wise to look for twice as much area as that required initially, because of the impact of increasing production volume in future

3. *Make inventory*: Prepare scale maps of farm area showing slope, utility map, building location, power generation lines drives, servicing and road rights of way and other important physical features. Review existing buildings, their usefulness, condition, locations and adaptability.

4. *Information*: Identity facilities needed, additional services required, family living desires, tax and insurance considerations.

5. *Organization:* Organize layout drawing and materials

A farmstead needs water, drainage, and production volume and off-farm factors. Careful planning includes reviewing the present, assessing the near future and providing for the more distance future. When planning a new building or adding to an existing farmstead, you must consider such things as:

- Site drainage
- Services (lanes, power, water supply, waste disposal)
- Security
- Space allowance for future expansion
- Separation distances for wind control, ventilation and disease control
- Distance separation from residences for control of noise and odours
- Municipal regulations
- Distance to wells, surface water, catch basins

Factors affecting farmstead planning

1. *Site selection:* Site selection is finding the best location available for the enterprise, enough space, asses to water, good drainage and other utilities and proximity to other farm operations

2. *Separating distance:* This factor help to locate one activity relative to other. Management requires that the distanced be close to the living center: pollution of air and water by some farm activities is not completely controlled by the present technologies. Environment of living areas should be properly protected from potential pollutants by keeping them at some distance. Nuisance including noise, dust, chaff, insects and heavy traffic. Locate annoying activities so that they do not depart from the safety standards and affect quality of living for the family.

3. *Topographic factor:* This is concerned with the layout of the land.

4. *Enough space:* This is needed for all farm activities and for expansion.

5. *Climatic factor:* Proper design and arrangement of farm buildings can help mires the bad effect of climate and take advantage of good effect.

6. *Service factor:* Services at the connecting limb between the farm stead building and facilities help the farm stead to operate efficiently.

7. *Fire safety factor:* Provide all-weather access to activity centers i.e. utilities, such as electricity, telephone water and sewage pipes. Fire prevention through adequate wiring, properly maintain heating equipments, lighting protection etc. Space buildings at minimum 16m apart to reduce fire spread and permit accesses for firefighting equipment. Natural hazard and accident prevention is part of the design considerations. Provide security to guide against theft and vandalism.

8. *Water supply and sewage disposal:* All things springs and are sustained by water. Municipal water supply systems generally comprises of collection or intake, purification or treatment, transmission works and distribution. Sewage disposal process involves solid waste storage, collection, transportation, treatment, utilization, processing and final disposal.

8.3 Farm building planning

In planning a farm building, adequate space must be allowed for each of the daily activities. This is not so much related to total space as it is to such things as door widths and heights, corridor widths, adequate space for a bed or a table and chairs, clearance for a door to swing open, etc. It is essential that these dimensions be checked in every design as very minimal changes can often make a considerable difference in convenience.

Various tribes and ethnic groups with different cultural and religious background have developed distinctive customs and social requirements. An analysis of the farm family's daily life, including present requirements and future plans, will help in selecting the important factors for designing an appropriate farm house.

Farm owners are responsible for obtaining a building permit for all agricultural construction projects, including manure storage facilities, grain bins and silos, along with all other farm structures. Whether building new or modifying an existing farm building, you must consider building code regulations.

The relationship of the farmhouse to the barn and other farm buildings is generally determined by such factors as: topography, weather conditions, convenience and labour efficiency, and most importantly for some settlers, ethnic or regional tradition. The following factors relevant to a farm house design are discussed as follows:

1. *Family size*: How many persons will live in the house initially and in the future? What are the family relationships - age, sex, marital status?
2. *Family comfort:* Are separate bedrooms and/ or houses needed for the husband and wife (wives)? Where do small children sleep - in parents' room, separate room nearby? Where do the older children sleep - separate room, separate house? Are children of different sexes separated?
3. *Resting/conversation:* What kind of room is required for resting and conversation - outside, verandah or separate shelter, - inside, kitchen or living room? Are men, women and children separated during these activities?
4. *Cooking/eating*: Is cooking done inside or outside the house or in a separate structure? Are cooking and eating done in the same area? Is there a separation between women and men, children or visitors during eating? What kinds of water resources are available?
5. *Store*: How much food should be stored, where? What types of storage conditions are required? What other items are needed to be stored e.g. fuel, water, implements?

Special requirements of farm houses

Additional factors must be considered in designing the farm house. They include the following:

1. A well-drained site, but suitable for a well, and when necessary either a latrine or a septic tank and drainage field. A home should never be built on a flood plain.

2. The relation of the dwelling to other farm buildings that will allow a view of the access road and the farmstead.

3. The correct orientation of the house to give protection against sun, rain, odour and dust while providing for ventilation, a view and easy access. An east-west orientation to provide the most shading is a general rule. However, it may sometimes be desirable to modify this to take advantage of a prevailing wind for better ventilation or to have more sun penetration into the house in cool highland areas.

4. A design which will allow building of the house in stages according to the availability of finances.

5. Flexibility in the arrangement of rooms to allow for alternative use and future expansion.

6. A kitchen large enough to allow for space-consuming activities such as cutting meat after slaughter, preparation of homegrown vegetables, etc.

7. A separate entrance from the backyard into the kitchen area. A small verandah at the rear where some of the kitchen works can be done and perhaps farm clothes can be stored.

8. A verandah large enough to allow for activities such as eating, resting, visiting, etc. The veranda, along with windows and ventilation openings, may need to be protected against insects with mosquito netting.

9. A separate office for larger farms, while a storage cupboard and the dining table will be sufficient for small farms.

10. A place to store dirty farm clothes and shoes combined with washing facilities, if possible. A room for guests if it is likely to be needed.

Building arrangement

When a site has been selected, it is then needful to draw a map which will show major details. There should be contour lines, the direction of the north, direction of prevailing wind and general slopes, existing roads, natural wind barriers and water ways. The arrangement and rearrangement of buildings should then follow till a satisfactory layout is designed (Omobowale, 2012).

An operation center should be located first, this will often be the farm house; the farm house should be sited such that it can be accessed from any direction in the farm stead, in general cases, it should be centrally located since the entire farm is administered from the farm house. The remaining buildings can then be arranged in relation to the operating center.

Building arrangement requires the consideration of some environmental factors such as slope, prevailing wind, sun etc. buildings should be located on relatively high ground with surface drainage directed away from foundations. Buildings should be arranged to take advantage of natural conditions; winds can blow in all direction but the prevailing direction is important, winds carry odours, dust, and noise, proper arrangement of buildings will use the wind to carry these away from the living center.

Livestock yards and buildings should be located downwind (wind ward) from farm home and from neighbours. Buildings lined up at right angles to the wind rather than parallel are less subject to the spread of fire.

Also, labour efficiency is improved by reducing travel to a minimum; buildings which will require frequent movement of workers should be sited close. Arrange buildings in relation to drive and yard to allow easy manoeuvring of large vehicles and equipment.

8.4   Farm workshop building planning

An efficient farm workshop is essential on every modern farm. The farm shop serves a multitude of roles, including equipment maintenance, fabrication and service, parts and tool storage, and as an office. Parts storage and service is becoming more important as service centers and parts distributors become more scattered and it may be necessary to travel many miles to get needed parts.

A shop provides a place to service, assemble, repair, adjust and modify equipment and keep tools in one location for field and farmstead operations. Overall, farm labor efficiency can be significantly improved by providing good shop facilities which will encourage preventive maintenance of equipment. Good preventive maintenance is known to significantly extend machine life and reduce the chances of costly downtime. It is also possible for the operator to construct or rebuild equipment to suit particular needs.

Location of farm workshop on farmstead

A producer will likely go in and out of the shop more times than any other building on the farmstead except the house. It should be located convenient to the house, machine storage and other farm traffic routes. It should be near the center of farm operations.

The shop should be at least 150 feet from the house to allow for easy equipment movement. This helps keep noise away from the living area and still provides for security. During the growing season, equipment will be moved and parked near the shop, which can detract from the appearance of the house and landscaping if too little room is provided. Machinery too close to the house can create a safety hazard if small children are around.

A shop should be situated on the farmstead so it is accessible only by passing near the farm home. Locate large doors so that valuable tools and equipment are not visible from the main road. Also, large doors should face away from the prevailing winter winds. In North Dakota, they should face south and/or east.

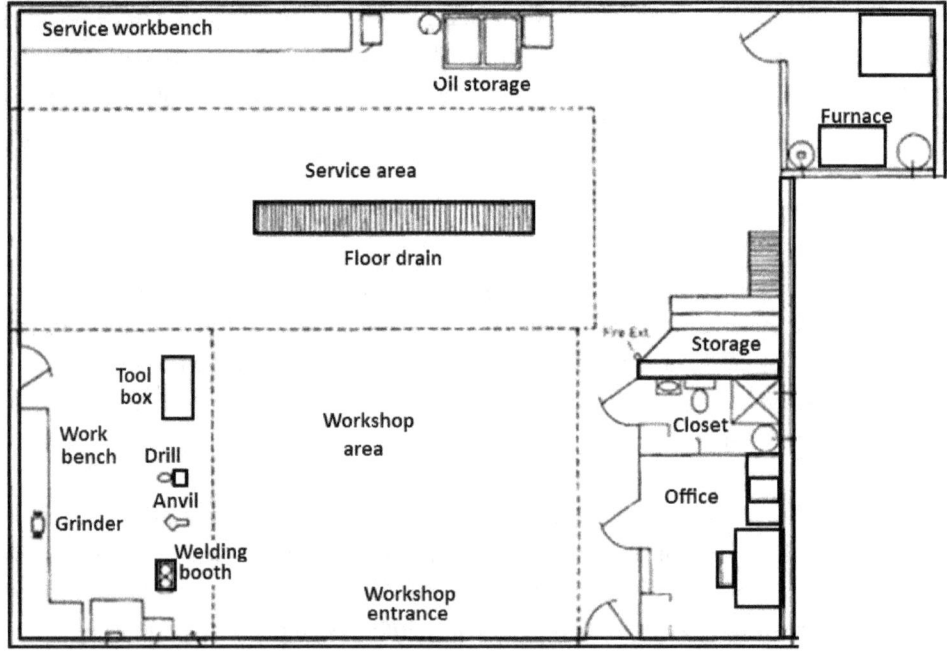

Figure 8-9: Workshop building plan

Good drainage is important. Shop floors should be at least 12 inches above existing grade and slope approach aprons 3 to 5 percent away from the shop.

Planning the farm shop building

Factors that need to be evaluated in a shop include: space requirements, equipment layout, access, storage, insulation, wiring, lighting and heating, ventilation, and office requirements.

The size and type of structure that houses the shop, the need for an outdoor work area and the kinds of tools, equipment and supplies that are used will vary according to individual needs. The ideal farm shop should include a large concrete slab outside where a farmer can park a machine for service or repair in warm weather. Tools and equipment should be conveniently located in the shop so equipment attachments can be installed or removed quickly.

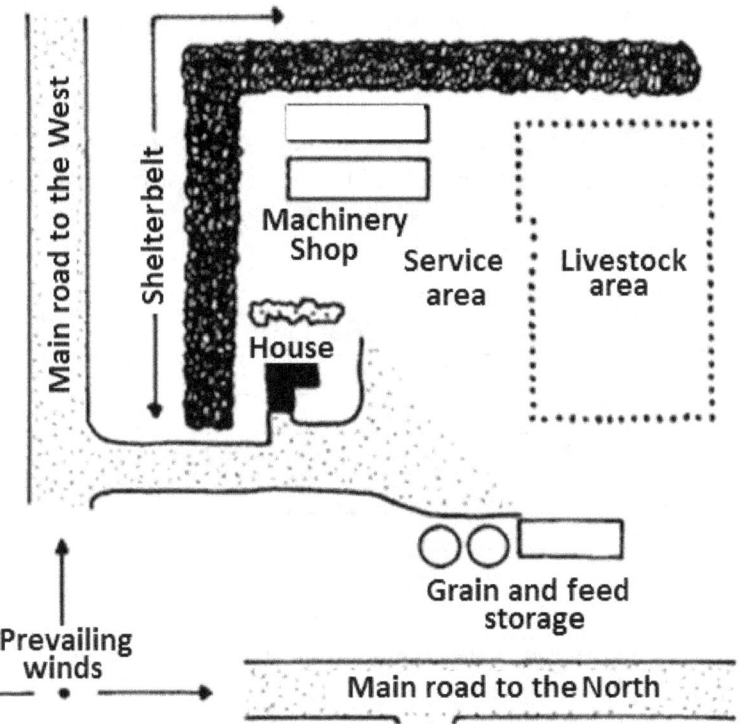

Figure 8-10: Farmstead layouts in relation to main road

Welding equipment should be located near the large door so repair jobs can be completed outside without having to move the welder or use extension cords with the welder. The ramp needs to be equipped with floodlights so machines can be serviced night or day. The inside area should be large enough to run most equipment inside for major repair jobs.

Suggested minimum shop size in relation to the size of the farm is shown in Table 8-1. The suggested building width should be at least twice the door width. The recommended door width should provide at least 2 feet of clearance for equipment being brought into the shop.

Table 8-1: Minimum shop dimension based upon farm size

Farm size (acres)	Shop size (sq. Feet)	Shop width (feet)	Shop length (feet)
1,000	1,680	40	42
2,000	2,000	40	50
3,000	2,300	46	50

Source: Hofman and Hellevang, 1994

A smaller service door is suggested so that the large door does not have to be opened for routine activities. The door height should provide at least 1 foot of clearance above machinery. Suggested minimum sizes are shown in Table 8-2.

Table 8-2: Suggested minimum farm shop door dimensions

Farm size (acres)	Door dimensions	
	Door width	Door height
1,000	20'	14'
2,000	24'	15'
3,000	28'	15'

Source: Hofman and Hellevang, 1994

A concrete apron to work on equipment outside the shop is encouraged. This provides a firm surface to park machinery on and reduces the amount of dirt carried into the shop on implement tires.

Table 8-3: Suggested dimensions of concrete apron outside shop

Farm size (acres)	Apron width (feet)	Apron length (feet)
1,000	20	40
2,000	25	40
3,000	30	46

Source: Hofman and Hellevang, 1994

All machinery access door frames should be protected both inside and out with concrete filled metal pipe at least 4 inches in diameter and 4 feet high. Place them in concrete 12 inches out and in line with the door frame or slightly inside the door.

Many older shops are remodeled farm buildings that may not be well suited to today's large farm equipment. Most existing farm structures are extremely difficult to convert to a modern shop because of the need to install a large door. An exception to this may be a relatively new machinery storage building where one end is often used for a shop. A concrete floor, ceiling and insulation can be added and a wall installed to partition off the machine storage area. An insulated overhead door may be needed and better lights installed to make the shop more usable.

Electrical service

An electrical service of 200-amp, 240-volt is the minimum recommended for all shops. In larger shops, a larger service may be needed. This will provide power to operate power tools, air compressor and welding equipment. Where motors 1/2-HP or larger induction motor will be used, install a separate circuit to serve not more than two 1/2-HP motors, and not more than one 1-HP or larger motor.

Install a 50-amp or larger 240-volt outlet for the welder. Two or more outlets properly located will make it possible to use the welder in most locations when working on machines inside the shop. Have at least one welder outlet near the large door so welding can be done outside.

Electrical lighting

Plenty of light in a farm shop is essential. Good lighting over the workbench is necessary for detailed work. Some natural lighting is however recommended. A window installed in a wall facing the entrance to the yard and small windows in the overhead doors allow workers in the shop to see arriving visitors. Windows in the door are also handy for those inside the shop to see when a door should be opened to drive equipment inside.

Fluorescent lighting is recommended in heated shops. In cold shops, fluorescent lights are extremely slow to start. Recommended lighting levels are shown in Table 8-4. If it is desired to use mercury vapor or high pressure sodium lamps for general lighting, install at the level of 0.8 watts per square foot for mercury vapor lamps and 0.4 watts per square foot for high pressure sodium lamps. These types of lights are slow to come on, so if mercury vapour or sodium lamps are installed a few incandescent lights may be helpful for initial lighting.

Table 8-4: Recommended lighting levels

	Foot Candles	Fluorescent lights (watts/sq.ft.)	Inflorescent lights (watts/sq.ft.)
General	30	½	2
Bench	50	1	4
Office	70	1 ½	6
Detailed work	100	2	8

Source: Hofman and Hellevang, 1994

For outside lighting, provide for 3.0 foot candles of illumination. Install a 300- to 400-watt photo-cell controlled mercury vapor lamp or 200 watt high pressure sodium lamp over the outside entrance about 20 feet above ground.

Ventilation

Ventilation is necessary to remove welding smoke and engine exhaust. A welding hood with a fan that will move about 1,000 to 2,000 cubic feet per minute for each welder that is being used is recommended over the welding area. If vehicles are to be operated in the shop for a period of time, an exhaust hose through the wall may be adequate or a small suction fan pulling exhaust through a tube from the exhaust pipe may be a better solution. Usually a fan capable of moving about 250 cfm per vehicle through the pipe should be adequate.

General ventilation in a shop is usually not needed. Opening and closing doors will usually provide adequate ventilation.

Heating system

The size and type of heating system will depend on the size of the shop and how often large doors will be opened and closed. Many types of heating systems are used in shops, including hot water pipes embedded in the floor, electric resistance cables in the floor, infrared and forced air furnaces. Most are powered by electricity, propane, coal, wood or fuel oil. Waste oil furnaces are becoming popular due to the difficulty in disposing of waste engine and transmission oil in rural areas. Several of these heaters are designed to burn fuel oil as well as waste oil. This provides an alternate fuel supply if waste oil supplies are not adequate.

Tool benches and parts storage

Keeping tools in place in a shop is often difficult. The advantages of having tools hanging on tool panels with the tool outlined include tools being easy to find and easy to put back in place, and one glance at the panel will tell what tools are missing. This is applicable in the temperate regions where temperatures are extremely low.

The arrangement of tools on the panel is a matter of individual taste. Generally, it works best to have the various types of tools grouped together. Extra tools can be stored in drawers built into the workbench.

Floor designs

If washing of machinery is to be done in the shop, a good drain system is a must. The simplest type of floor drain is to slope the floor to the center and to the doorway. A center floor drain with an oil and sludge collector is a better option but is more difficult and expensive to install. An example of this type of drain is shown in Figure 8-9.

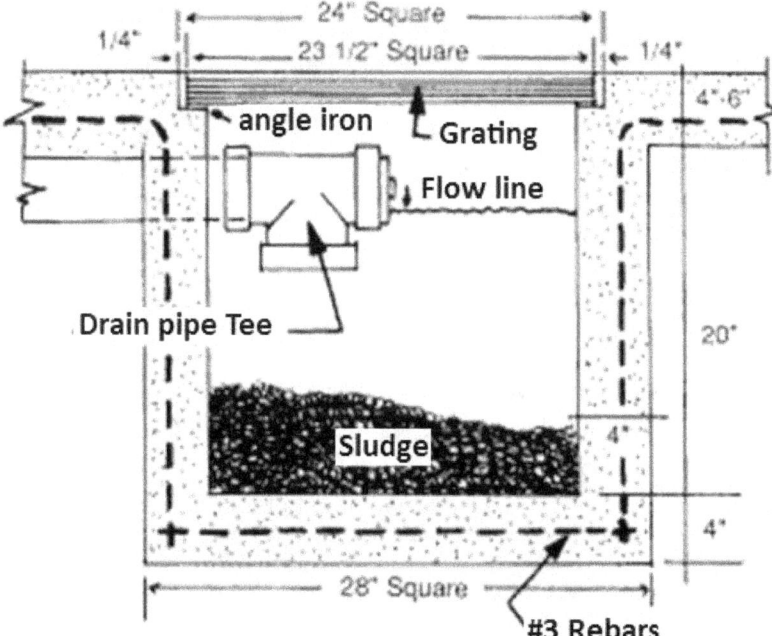

Figure 8-11: Floor drain

The drain is designed to catch oil on top of the water and collect heavy solids in the bottom of the pit. This type of drain will collect these materials before they flow into the drain field as long as it is cleaned on a regular basis.

Another type of drain that works well in the service bay is a long narrow type which slopes 0.1 inch per foot of length.

Office

The extent of the office will vary considerably depending on need. A minimum recommendation is a desk, telephone and storage for service manuals and records. A restroom with running water and shower is optional, along with a washing machine for greasy or pesticide contaminated clothing. It is convenient to have an exterior entrance into the office, and this provides a good entrance for farm visitors. The office will usually need additional climate control such as baseboard heating panels.

8.5 Farm roads and construction

Farm roads

Road transportation was primarily a means of communication from the creation of man. Pathways are the original form of roads available before man stated the use animals. Then with the use of animals, animal drawn vehicles became important.

Farm road history

During the Roman Empire the real science had three classes of road:

1. *Level or earth road,*
2. *Gravel surface road and*
3. *Stone block roads.*

With the inventions of wheel vehicles man sees the necessity of road. The first earth surface road was constructed in 3500BC in Mesopotamia after the discovery of wheel vehicles. A stone surface road in the Island of Crete was dated 1500BC. The Bible also mentioned the construction of road from Babylon to Egypt in 539BC.

Classification of roads

Classification may be based on road design (construction) road, use (traffic characteristics), or function of which the road is to be served. Three criteria are basically important for road classification. These are:

- *Traffic volume* to indicate the number of vehicles plying the road.

- *Tonnage load transport*: This is likened to the population density. It is important in the industrialized areas
- *Location and road function*: This classification is base on the location and the function the road is to serve.

For planning purposes, a classification by function is most useful. Classifications by road design are commonly based on pavement width and type. Road surfaces can be paved with seals or oil-based materials as well as asphalt or concrete. Unpaved roads are characterized as earth, sand, or gravel. Road traffic characteristics embody features such as traffic volume and composition by mode.

Classifications based on tonnage loads consist of high density, low density and light density and these are called relative roads. Under this classification five basic roads are available.

- *National road or highways*: These roads connect ports, capitals, foreign countries and cities. They must have at least two lanes. The width is 8m and the shoulder is 2m wide.
- *State roads* connect cities with the national highways. They have two lanes and a width of 8m, 2m shoulders.
- *Major district roads* connect cities or local governments.
- *Minor district road* connects town and important village to the cities.
- *Rural (feeder) roads*: These are the roads mostly of great concern to agricultural engineers. It links towns, village and market centers. Erosion is the major problem on these roads and is often called farm roads.

Farm roads

Farm roads are specially designed for farm operations and rural settlements. They include all access roads to the farm and within. They are mostly made of consolidated earth materials, gravel, concrete materials, farming materials in order to give required surface conditions for proper traction.

In USA, Sweden and Japan the percentage of the earth road are 18%, 38% and 52% respectively. It is much higher in central and south America, Nicaragua, Brazil, Argentina, Mexico and Australia which have 93.3%, 93.6%, 73%, 55%, 53% and 71% respectively.

Figure 8-12: Farm road linking Ishiagu and Mile 2 farm community

The ride-ability of earth road is contingent on the type and condition of the soil making up the road. A particular soil characteristic is a function of climate. Season also affect the performance of farm roads. The moisture content of the soil surface increases during raining season and there is sliding of the wheel of the equipment hence the need for drainage to arrest the situation.

Traction and farm roads

Speed is not the major design criterion for farm roads rather the traction is the basic design criterion. There must be a very good grip between the wheel and road surface and the traction must be effective. Surface friction must however be minimized in order to prevent excessive wearing of the wheel.

Farm roads and vehicle draft

The draft (ability to pull) must be reduced (of the wheel) and it could be noted that the draft is a function of axle friction, rolling resistance and slope resistance. So the use of high grade ball or roller bearing together with some lubrication has effectively minimized rear axle friction. Low value of rear axle friction is recommended for effective traction otherwise higher value of axle friction results to higher power requirement.

Farm roads and vehicle rolling resistance

Low rolling resistance is usually recommended for farm vehicles, it should be noted that rolling resistance is mostly affected by wheel diameter and the weight of the machine. Hence large wheel diameter and heavy wheel are associated with high rolling resistance, which is undesirable.

Pneumatic types are therefore recommended for farm vehicles on soft ground or earthen surface to ensure a low rolling resistance because pneumatic tyres have lighter weights. High graded slopes are not recommended on farm roads since the resistance to traction is proportional to slope. In theoretical analysis of wheel rolling on bare soil, it is normally assumed that the resistance, which the soil offers to the wheel, varied with the penetration depth z.

$$q = C \left(\frac{z}{z_0} \right) \text{............................} 8.1$$

Where

- q = Soil resistance
- Z = Depth from the surface
- Z_0 = Depth of the surface to the datum
- C = Soil resistance at depth Z_0. It depends on the moisture content and the impermeability of the soil.

Bousineq formula could be used to obtain the soil stress as follows if a circular plate test is used.

$$\sigma_{zo} = \mu \rho \left(1 - Cos\mu\alpha \right) \text{........................} 8.2$$

Where

σ_{zo} = Vertical at stress at depth Z below the center of circular plate.

ρ = Unit loads over the area.

μ = Stress concentration coefficient. A value of 3 is assigned for solids and 6 is recommended for plastics.

α = Angle at which the plate diameter is seen from measurement and by calculation.

$$\alpha = 2tan^{-1} \left(\frac{r}{z} \right) \text{........................} 8.3$$

Where

r = Radius of plate

Causes of road failure

Farm roads fail as a result of one or a combination of some of these factors:

a. Weathering,
b. Organic growth,
c. Wears,
d. Poor drainage,
e. Deterioration due to effect of aging, material failure and
f. Designed construction fault.

Maintenance of farm roads

This is purposely to offset the effect of those factors responsible for road failure

It involves physical maintenance activities such as patching, filling of joints and cracks on the roads, grading, compaction, clearing and so on. This is to ensure that the road is save and in convenient condition for economic use and easy and effective movement of vehicles.

The maintenance of drainage system is of a vital importance in the management of farm roads.

Maintenance of road could be classified into three categories thus:

- *Normal repairs*: This includes minor repairs and maintenance e.g. patching, filling, regular check up and inspection of roads.
- *Repair of bridge, culvert and drainage system*: This is a major maintenance to make bridge and culverts through for the passage of transport vehicles and runoff water. Adequate drainage must be provided for a farm road, shoulder and edge of the road to convey water away from the road surface.
- *General maintenance*: This is the normal repairs and maintenance but is more general e.g. clearing the roads of trees, bushes, stays etc. the road surface should be maintained e.g. filling of potholes, removal of broken parts of automobiles after an accident etc.

Farm transport standards and specifications

Standards and specifications are essential aspects of engineering design and construction. Specifications are written instructions, which set the standards, and are based on the results of experiments and research knowledge acquired over many years relating to the quality of material and the workmanship, which can be demanded and expected on a particular type of scheme or project. It is most important that the specification should describe every constructional item, which enters into the scheme or contract. The material to be used must be specified together with the quality they must meet.

For transportation; standard specifications are mostly available with highway authority such as Ministry of Transport (MOT) and are usually printed in permanent book forms. However, situations arise such that particular problems arise which may necessitate a deviation from the existing standard specifications, hence the final specifications for such project should include supplementary provision that covers the condition peculiar to the project.

Specifications are integral parts of construction documents and their preparation should not be taken lightly. In preparing specifications, legal implications must be seriously considered. Legal opinion should be sought before specifications are placed for contract. In preparation of specification for farm road the following features must be given serious consideration.

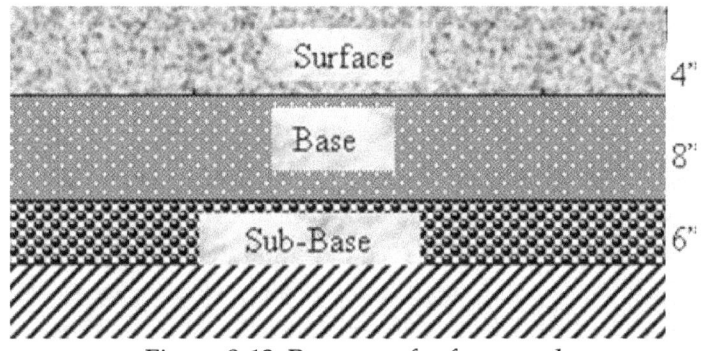

Figure 8-13: Pavement for farm road

The pavement: Farm roads mostly have flexible pavement with a relatively thin layer of tar or bitumen bond materials or a lean layer of concrete under which there is often a sub-base of granular materials to or stabilizing soil.

Surfacing: The surfacing of farm road is mostly done with stabilized earth material. Occasionally thin layer of gravel will always be required for surfacing to increase friction and enhance surface drainage of water. The surfacing of road is mostly made up of two courses namely the wearing course and the sub-base course. The wearing course is always thin and made up of a more stable material.

Drainage, moisture control and soil stabilization: Water drainage from road surface is very essential to avoid road failure and promote load-bearing properties of soil and also increase road durability. A suitably graded soil material with an adequate cohesion and internal friction should be used as construction material for farm road. Cement, gravel-sand-clay, and natural deposits are good examples. Well-graded soils permit a high dry density, compaction with low pore spaces hence the risk of subsequent increase of moisture content is reduced. Surface moisture could be controlled using the followings; Lime stabilization, Bitumen stabilization or other chemical agents.

Bridges and culverts: Bridges and culverts are required for the drainage of surface water. Road surface water could be generated as a result of; precipitation, flood from running stream or river drainage, intercepting water course or seepage.

The following measures can be taken to effect drainage.

- Drain culverts should be provided to control surface water. Adequate design of hydraulic structures must be done.
- Provide slating side shower.
- Provide roads with verge raised above the carriage.

8.6   Agricultural wastes management

Wastes

"Wastes" generally includes any solid or suspended materials, dissolved or transported in water (including sediment) which are spilled or deposited on land or into a water resource in such volume, composition or manner as to cause, or to be reasonably likely to cause pollution. It is any matter, whether gaseous, liquid or solid or any combination thereof, which is from time to time designated as any undesirable or superfluous by-product, emission, residue or remainder of any process or activity.

A common example is abattoir waste which can be defined as waste or waste water from an abattoir which could consist of pollutants such as animal faeces, blood, fat, animal trimmings, paunch content and urine.

Types of farm wastes and their management

1. Agricultural wastes

Common agricultural wastes include:

a.   Packaging materials;
b.   Silage plastics;
c.   Redundant machinery;
d.   Used tyres;

e. Used oils;
f. Used batteries;
g. Old fencing materials;
h. Scrap metal; and
i. Building waste.

The types and quantities of wastes vary between farms. Other less common wastes include unused pesticides and veterinary medicines, horticultural plastics and spent sheep dip. The proper management of waste from agricultural operations can contribute in a significant way to farm operations.

Waste management helps maintain a healthy environment for farm animals and can reduce the need for commercial fertilizers while providing other nutrients needed for crop production. Agricultural waste typically associated with animals includes but is not limited to manure, bedding and litter, wasted feed, runoff from feedlots and holding areas, and wastewater from buildings like dairy parlors.

Agricultural wastes management

Farm waste management covers the responsible storage, collection and disposal of all farm waste and the preparation and implementation of a farm waste management plan. Responsible farm waste management aims to improve the quality of our watercourses beyond a level that is required by current legislation and GFP and will improve the visual appearance of the farm and farmyard.

The farm waste management plan must take into account the collection, storage and disposal of all farm wastes. Implementation of the plan will reduce the risk of pollution and prevent the loss of valuable nutrients in slurry and farmyard manure.

2. Wood wastes

Wood waste represents a tremendous volume of fiber which is yet to be fully realized as a viable source of secondary fiber. One of the fundamental distinctions made with regard to wood waste streams is between "green" wood and "dry" wood (CWC, 1997). "Green" wood refers to fiber that has not been dried, while "dry" wood refers to fiber that has been dried. Green wood typically has moisture content by weight of 50 percent or more (wet basis), contrasted with dry wood which typically has a moisture content of 10 to 25 percent by weight (wet basis) (CWC, 1997). Waste chips include wood chips of irregular size (fines) and may be contaminated by bark or rot. Waste chips have relatively high moisture content, about 50 percent.

Wood wastes management

Wood waste processing consists primarily of sorting, size reduction, and then screening for contaminant removal and sizing. The elements must all work together in order to form a successful wood processing operation. Innovative processing employing size reduction equipment (usually hogs) with ancillary feeding, screening and cleaning equipment have allowed dramatic improvements in the ability to remove contamination from wood waste. Contaminant removal has continued to be keyed to the successful high grading of low value wood wastes.

3. Animal wastes

Animal waste includes poultry droppings, feed wastes, cow dung, and all other disposable poultry handling materials.

Animal waste management

Protecting the public's health and safety, as well as the Aquifer and the public's drinking water supply are the primary objectives of the solid and liquid waste management programs, operated by the Environmental Health division. Proper waste handling helps prevent pollution and problems with animals and insects.

Manure management is a critical concern in many parts of the world where the expanding livestock and poultry industry is concentrating. In past literature manure and related animal-production residues often have been referred to or classified as "waste." In this presentation, the term waste is replaced by *manure* and *residues* as much as possible in an effort to maintain the concept that manure is a resource to be used rather than a product to be wasted.

Animal manure contains a number of contaminants that can adversely affect surface and groundwater. The principal constituents of animal manure that impacts surface and ground water are the organic matter, nutrients, and fecal bacteria. Manure may also increase the amount of suspended material in the water and affect the color either directly through the manure itself or indirectly through the production of algae.

Indirect effects on surface water can also occur if sediment enters streams from feedlots or overgrazed pastures and from eroded stream banks at unprotected cattle crossings. In a natural environment the breakdown of organic matter is a function of complex, interrelated, and mixed biological populations. However, the organisms principally responsible for the decomposition process are bacteria. The size of the bacterial community depends on its food supply and other environmental factors including temperature and pH.

Methods of handling animal waste

A complete waste disposal system is not a luxury in livestock business but a necessity. Careful waste management is needed to:

a. Maintain good animal health through sanitary facility.
b. Avoid pollution of air and water.
c. Compliance with Local, State and Federal regulation.

4. Solid wastes

Solid waste includes garbage, rubbish, ashes, industrial wastes, demolition and construction wastes, abandoned vehicles or parts, discarded items, wood waste, medical waste, tires, recyclable items, sewage sludge and seepage (from septic tanks).

Handling solid manure

Solid manure results from packing and holding excrement in bedding or by allowing the liquid to drain off. Handling of solid manure requires;

a. Solid floors that can be bedded or drained.
b. Minimum equipment required.
c. An area on which to spread the solid.

Suggestions on waste handling

1. Install sloping floors; locate water where manure accumulation is deserved.
2. Hull manure directly to fields whenever possible but avoid spreading on frozen fields.
3. Locate storage in convenient place to ease loading with spreader. Only solid manure can be spread by this type of equipment (Figure 8-14), because the percentage of dry matter must be at least 13%. Between 8% and 13% we find manure with small quantities of straw, which requires same type of equipment but equipped with a watertight tank.
4. Divert surface water away from storage area.
5. Control runoff from stockpiles.

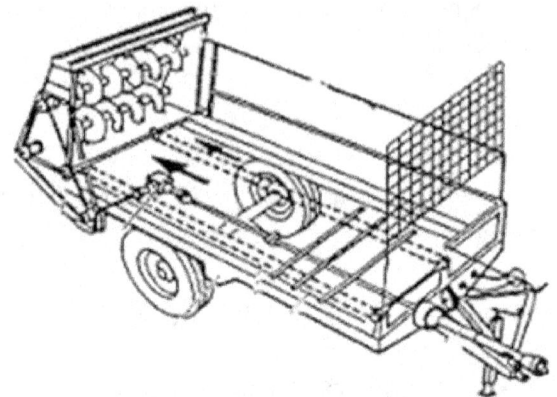

Figure 8-14: Solid manure spreader

5. *Liquid wastes*

Liquid manure is waste water effluents from agricultural and industrial processes. Examples of liquid manure include waste water from treatment plants, poultry wastes etc.

Liquid wastes/manure management

Liquid manure may have to be spread on soils covered with snow or when it rains strongly. Liquid manure handling requires the following facilities and equipment.

a. Scrappers, gutters, slotted in storage or treatment.
b. Storage treatment unit to which water can be added.
c. Pumps, agitators and augers to stir and remove the manure.
d. Vacuum tankers, truck or wagon irrigation equipment, (Figure 8-13) is used for spreading slurry or liquid manure with no more than 8% dry matter. Such tankers may be equipped with systems allowing one to bury liquid manure in the soil; the advantages of burying are both avoiding losses of NH_3 in the air and increasing spread able land, because there are no odors while spreading and so it may be done nearer to dwelling areas.

Figure 8-14: Vacuum tanker

1. A lagoon can treat wastes before field spreading.
2. Liquid manure may be spread with *irrigation systems* equipment similar to those used for irrigation, in order to avoid evaporation, if it is much diluted (less than 3%).

Collection of livestock wastes

Several collection methods for livestock wastes are possible with some handling systems. Some system combine collection and storage function such as a built-up manure park of slotted floors over a liquid tank.

In selecting a collection system, consider the following factors

1. Type of production system.
2. Type of facility
3. Labour requirement
4. Investments
5. Total volume of waste handled.

Structural requirement for livestock wastes collection

1. *Slotted floors:* These types of equipment / structure provide rapid separation of an animal from its manure. Size of opening and space between openings depend on the manure, its property and on expenses. Concrete slots are perhaps the most common and durable, but are heaviest and require strong supports wood slots can wear out, warp thus leaving irregular slot spacing. Metal and plastic slots are more uniform and easier to handle, install and replace than concrete, though expensive.
2. *Scrappers and blades*: these are hand held tools that can be used to collect manure. They are used on solid floors or under slotted floors, in gutters, channels or troughs to move wastes to storage, treatment or spreading equipment.
3. *Front-end loaders*: conventional agricultural tractors mounted with loaders are available in variety of styles and sizes. Front end loaders remove solid manure from open slots, storage or building floors. Other farm uses include loading silage, bedding, fertilizer or other bulk materials.

Animal waste utilization

Regardless of the storage treatment and handling method used on animal feeds, some end products remain. These remains are valuable resources to be utilized for benefit or returns. The end use of waste often dictates the waste disposal system. Animal manure is most commonly used for fertilizer and soil conditioner.

Direct disposal of animal waste is desirable because

1. The nutrient helps to build and maintain soil activities.

2. It increases water-holding capacity,

3. Lessen wind and water erosion,

4. Improve soil aeration and promote the growth of beneficial organisms.

The economic value of manure fertilizer is calculated from its available NPK content. Wrong application can be harmful to crop, soil surface and ground water quality.

Livestock waste storage

The design of livestock housing is often tied closely to selection of a type and size of waste storage. A storage unit may be an outdoor tank, silo or pond or park. Size of storage depends on farm requirement, climate, and the need for pollution control and minimum size of waste to be handled. Ventilation in waste storage is necessary because of primary hazard from manure gases.

Maximum building ventilation is required when agitating or pumping wastes from a pit. Accumulated air in storage should be replaced in the fresh air through open doors, windows and large fans. Alarm systems (loud bells or light) to warn of power failures in totally enclosed building is essential.

Waste management practices used to collect and store animal wastes include a variety of scraping and flushing systems and storage structures such as tanks, lagoons, ponds, and sheds. The choice of a collection method and storage structure depends, in part, on the volume and moisture content of the waste being handled. For example, wastes with relatively high moisture content, such as dairy and hog waste are suitable for a mechanized scraping or water-based flushing system. In contrast, drier wastes, such as beef cattle and poultry waste are typically moved with a tractor or through manual labour.

Choice of a storage structure depends on waste volume and moisture content. Structures such as lagoons, retention ponds, and tanks are suitable for very wet waste, such as waste slurry. Structures such as sheds or synthetic covers are used for dry wastes such as poultry litter.

Location of waste storage

Avoid locating storage on creviced bedrock, gravel beds or below the water table; if shallow bedrock is present, obtain special review by qualified engineer or contact health boards. A distance of at least 30m is provided between source of water supply and the nearest part of storage. Storage should be shielded from public roads and family living area.

Maintenance of waste storage

Periodic inspection should be made for the tank and its surroundings for leaks, deterioration of grills, covers and ladders. A vertical distance from ground to lid of about 48 (46cm) discourages traffic.

Best waste management practices

A wide variety of animal waste management practices are currently available to livestock and poultry producers. A farmer's selection of a particular practice or system of practices depends on site-specific factors—the type and volume of waste to be managed and the proximity of the production facility to surface water or groundwater—cost considerations, and state and local regulations.

These practices include techniques to:

1. Limit waste runoff, such as cementing and curbing animal confinement areas or planting grassed buffers around these areas.
2. Collect and store waste, such as scraping or flushing systems and storage tanks or retention ponds.
3. Alter or treat waste, such as reformulating feed mixes or composting.
4. Use waste, such as an organic fertilizer, an additive to animal feed or on-farm energy generation.

Municipal waste water management system

Municipal waste water system structures comprises of a sewage or drainage system. Domestic waste waters are from house-holds, manufacturing industries and commercial establishments referred to as industrial or trade wastes, in free or gravity flow, sewers and drains flow consciously downhill except where pumping stations lift flows through force mains into higher lying conduits. They are not intended to flow under pressure but future provision included pressurized pipes in case of clogging of wastewater. Hydraulically, gravity sewers are disputed as open channel flow partly full or at most just filled.

In well watered regions along the seas and oceans, collected waters are usually discharged into the receiving bodies of water after suitable treatment. This is referred to as disposal by dilution. In semi arid regions, terminal discharge may be on to the land irrigation. Treatment before disposal aims at removal of unsightly putriscible matters, stabilization of degradable substances, removal of nutrient and minerals and destruction of disease-causing organisms all in a table degree. Biodegradable material as well as the nutrient in natural runoff add to the recurring waters and may produce massive algae blooms in lakes.

8.7 Wastewater treatment lagoon

Lagoons are treatment structure for agricultural wastes. Wastewater treatment lagoons are earthen impoundments that are engineered and constructed to treat as well as temporarily store wastewater. In practice, the terms lagoons and ponds are used interchangeably. The wastewater treatment lagoons are different from wastewater storage or holding lagoons in that they are designed to function as biological reactors that allow effective degradation of organic compounds contained in the wastewater by various microorganisms.

Therefore, the physical, chemical, and biological environments in the treatment lagoons are controlled to achieve the intended purposes of wastewater treatment. Lagoons contain nutrients, salts, and other soluble chemicals that, in many cases, are eventually applied to crops as fertilizer. While waste is stored and treated in the lagoons, seepage losses from the sides and bottom of the containment could potentially affect soil and ground water quality (Ham *et al*, 1999). Lagoons and retention ponds can be lined with packed clay or a synthetic material to minimize the leaching of liquid waste into groundwater.

Wastewater treatment lagoons have been widely used for the treatment of human, industrial, and animal wastewaters due to their low capital costs and simple operational and maintenance requirements compared with other biological treatment systems. Wastewater treatment lagoons range in depth from shallow to deep and often are categorized by their mode of biodegradation, as determined by the presence or absence of dissolved oxygen, source of oxygen, and other design features.

Lagoon classification

Based on the presence of oxygen, lagoons are classified as aerobic, anaerobic or facultative depending on their loading and design and can be used in series to produce a high quality effluent. Bacteria are the primary microorganisms responsible for waste degradation in all types of lagoons.

Aerobic lagoons

Aerobic lagoons contain dissolved oxygen in the water to sustain aerobic bacteria. The dissolved oxygen can be supplied naturally or artificially. Natural aeration is achieved by air diffusion at the water surface, by wind- or thermal gradient-induced mixing, and by photosynthesis. The photosynthetic microorganisms include algae and cyanobacteria (blue-green algae). Artificial aeration is achieved by mechanical aeration. Thus, there are two types of aerobic lagoons, naturally aerated lagoons and mechanically aerated lagoons.

The general chemical reaction for aerobic degradation of organic compounds is as follows:

$$(C, H, O, N, S + O_2 \rightarrow CO_2 + H_2O + NH_4^+ (or\ NO_3^-) + S(or\ SO_4^{2-}) \ldots\ldots\ldots\ldots.8.4$$
Organic compounds

Naturally aerated lagoons are quite shallow, typically 1 to 2 feet, to allow sunlight to penetrate the full lagoon depth to maintain active algal photosynthetic activity during daylight hours. The oxygen produced from the photosynthesis process is used by aerobic bacteria to degrade the organic waste.

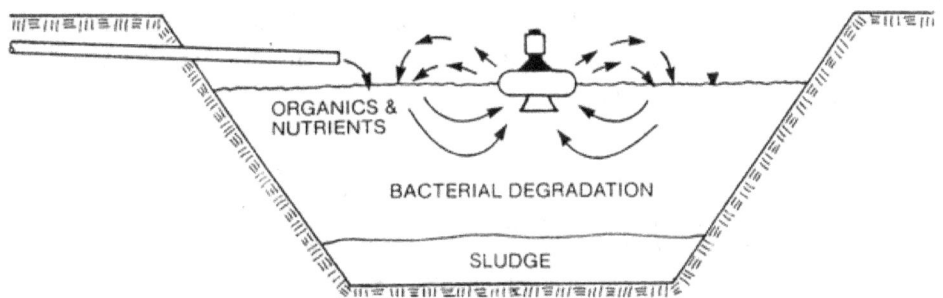

Figure 8-15: Mechanically aerated lagoons

Mechanically aerated lagoons (Figure 8-15), however, do not have the depth requirement. They are usually built with much more depth and a smaller surface area than the naturally aerated lagoons. Since oxygen is supplied through mechanical means, algal photosynthesis in the mechanically aerated lagoons plays an insignificant role.

Anaerobic lagoons

Anaerobic lagoons are used mostly for high-strength wastewater treatment, such as animal wastewater. Anaerobic lagoons are used to collect, treat, and store waste at concentrated animal operations (CAOs) centers. Anaerobic lagoons handle high loading waste but give off septic odours. They vary in depth from 8 to 30 feet and are built as deep as the local geography allows minimizing the surface area and reducing odour emissions. The top layer may contain dissolved oxygen depending on wind, temperature, and organic loading rate.

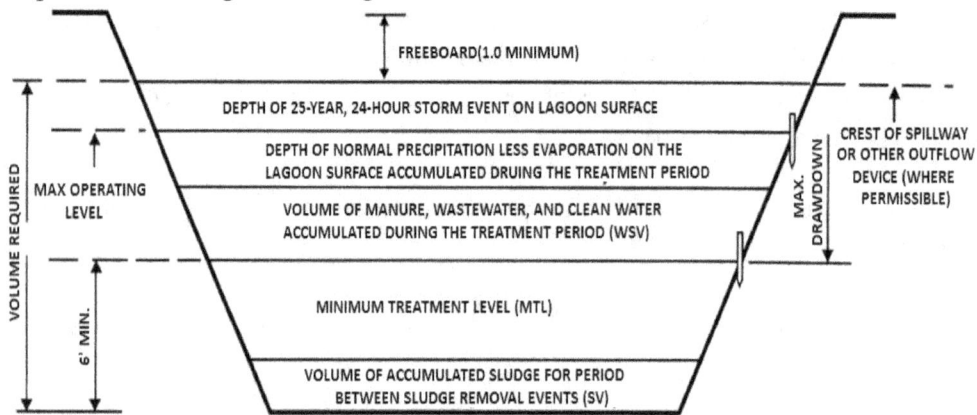

Figure 8-15: Anaerobic lagoon cross section (adapted from USDA-NRCS, 1992)

Under anaerobic conditions, two distinct reactions occur. In stage one, hydrolysis of organic compounds and conversion to intermediate organic acids are achieved by acid-forming bacteria called acidogens. Then in stage two, the organic acids are converted by methane and carbon dioxide by methane-forming bacteria called methanogens as illustrated in Figure 8-16.

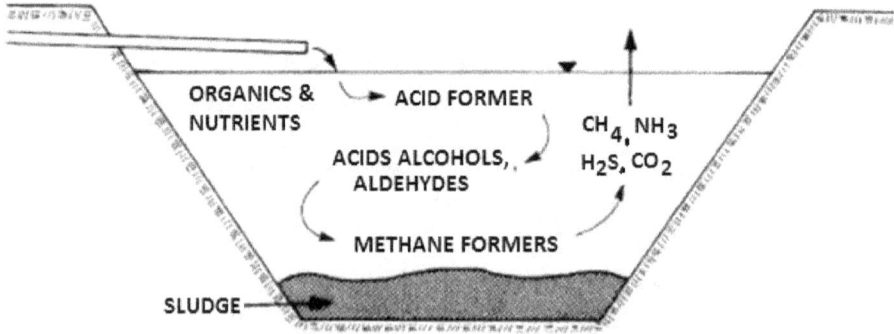

Figure 8-16: Degradation of organic compounds in anaerobic lagoons

The overall complete reaction of anaerobic degradation is:

$$Organics(C, H, O, N, S)CH_4 + CO_2 + NH_4^+ + H_2S) \dots\dots\dots\dots\dots\dots\dots.8.5$$

In most anaerobic lagoons, anaerobic degradation is not complete due to the fact that conditions such as organic loading rate, temperature, and retention time are not optimum for bacterial reactions. High concentrations of intermediate degradation compounds, such as organic acids, amino acids, aldehydes, sulfides and others are present in the lagoon water and contribute most of the foul odors. Well-designed and operated lagoons can effectively lower the concentrations of these odorous compounds and keep odors to a minimum.

However, high emissions of methane from open lagoons may be expected. Gases produced at the bottom of anaerobic lagoons often lift sludge to the top surface forming a layer of floating solids. Sludge accumulation as a result of incomplete degradation of solid waste depends on management, environment and waste characteristics because complete treatment is not practicable in anaerobic lagoon; they are not used in houses.

Facultative lagoons

The facultative lagoons are deeper than aerobic lagoons, varying in depth from 5 to 8 feet. Waste is treated by bacterial action occurring in an upper aerobic layer, a facultative middle layer, and a lower anaerobic layer. Aerobic bacteria degrade the waste in the upper layer where oxygen is provided by natural surface aeration and algal photosynthesis. Settleable solids are deposited on the lagoon bottom and degraded by anaerobic bacteria. The facultative bacteria in the middle layer degrade the waste aerobically whenever dissolved oxygen is present and anaerobically otherwise.

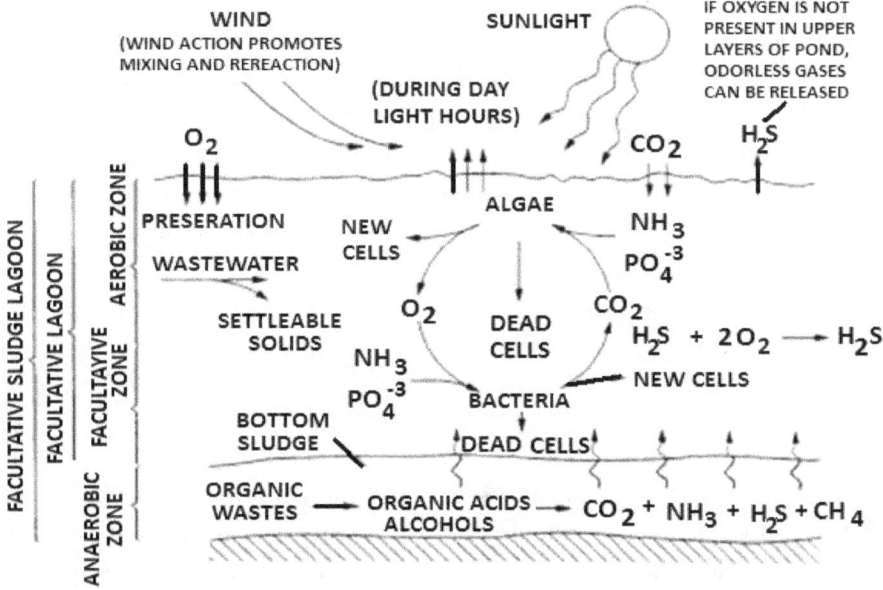

Figure 8-17: Facultative lagoons (adapted from Crites and Tchobanoglous, 1998)

Figure 8-17 shows microbial interactions and waste degradation pathways in a facultative lagoon. The facultative lagoons are more common than naturally aerated lagoons. They have more depth and smaller surface areas but still have good odor control capabilities because of the presence of the upper aerobic layer, where odorous compounds such as sulfides produced by the anaerobic degradation in the lower layer, are oxidized before emission into the atmosphere. Biochemical reactions in the facultative lagoons are a combination of aerobic and anaerobic degradation reactions.

Environmental impact assessment (EIA) assessment of lagoon

A comprehensive environmental impact assessment of lagoons requires three focal areas:

1. *Toxicity* – What are the constituents in the lagoon waste that pose a threat to water quality and public health?
2. *Input loading* – At what rate does waste seep from a lagoon under field conditions? and
3. *Aquifer vulnerability* – How do soil properties, geology, and water table depth affect the risk of waste movement from the lagoon to the ground water?

Location consideration of lagoons

The following considerations are required for an effective lagoon establishment:

1. Lagoons should be located far from the farm houses; they are often located near the waste sources over impervious soil. Lagoons functions best above 70°f and bacteria activities nearly stops at freezing.
2. A lagoon requires stable, level ground where beams and structures can be built.
3. All proposed lagoons should be lined with an impermeable membrane.
4. Soil testing would be needed to confirm how the lagoon should be designed and constructed.
5. Lagoon design volume is a function of; number of animals and the average weight of each animal.

Earthen embankments for animal liquid-waste containment

Earthen embankments, which can be used to contain water for a variety of purposes, may also be used to construct animal liquid-waste storage lagoons. A properly designed facility, whether for animal waste or process water, will have adequate storage capacity for the intended use and for any surface water run- 25-year, 24-hour design storm, if applicable. Earthen embankments should be constructed in such a manner that the embankment will not fail, and — in the case of animal liquid waste storage — be constructed using materials and practices that will assure minimal seepage from the sides and bottom of the lagoon.

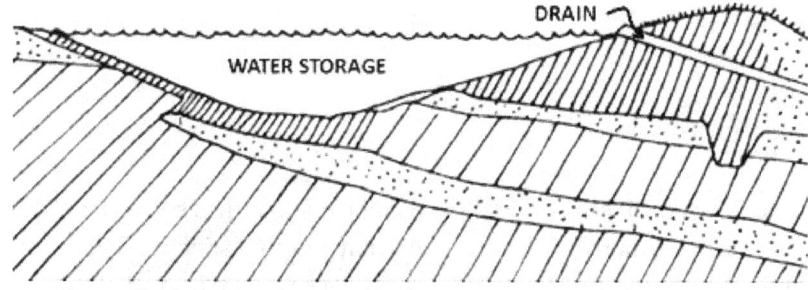

Figure 7-18: An earthen embankments water storage structure

Further reading

Ruihong Zhang, 2001. Biology and Engineering of Animal Wastewater Lagoons. University of California Davis

ASAE, 1999. Design of anaerobic lagoons for animal waste management. ASAE EP 403.3. ASAE Standards 1999. American Society of Agricultural Engineers. 2950 Niles Road, St. Joseph, MI.

Crites, R., and G. Tchobanoglous. 1998. Small and Decentralized Wastewater Management Systems. McGraw-Hill.

Gilley, J.E., D.P. Spare, R.K. Koelsch, D.D. Schulte, P.S. Miller, and A.M. Parkhurst. 2000. Copper and zinc in swine diets affect phototropic anaerobic lagoons. In: Proceedings of 8th International Symposium of Animal, Agricultural and Food Processing Wastes. October 9-11, Des Moines, IA. pp. 656-663.

Planning the Farmstead Distribution System, 1988. Notes taken from chapter 5 of Gustafson, Fundamentals in Electricity for Agriculture,

Sobowale A., 2012. Farm Structures. Lecture note on AGE 507. Date modified 10/06/2012

USDA-NRCS. 1992. Agricultural Waste Management Field Handbook.

Vern Hofman and Kenneth Hellevang, PE, (1994). Planning Farm Shops. County Commissions, North Dakota State University and U.S. Department of Agriculture cooperating. AE-1066,

CHAPTER 9

Farm Storage Structures

9.   Introduction

According to a local proverb, a rich person is the one that has stored enough grains to last the year round. Another proverb adds that the foolishness of the one who has enough grains in the house is taken as intelligence. Therefore, the purpose of any grain storage facility is to provide safe storage conditions for the grain in order to prevent grain loss caused by adverse weather, moisture, rodents, birds, insects and micro-organisms like fungi. Grain storage is therefore highly important for food security – both at local and national levels.

Figure 9-1: Grain storage structures

In many parts of Africa some crops can be produced throughout the year, while other food crops such as cereal grains and tubers, including potatoes, are normally seasonal crops. Consequently the food produced in one harvest period, which may last for only a few weeks, must be stored for gradual consumption until the next harvest while seeds must be kept for the next season's crop. Therefore the principal aim in any storage system must be to maintain the crop in prime condition for as long as it is possible.

Storage of food crops

Food crops fall into two broad categories namely;

1. Perishable crops and
2. Non-perishable crops.

This categorization is based on the rate at which a crop deteriorates after harvest and the length of time it can be stored. Cereal grains can be stored for over a year and are considered to be non-perishable, whereas tomatoes are perishable crops and when picked fresh, will deteriorate in days. Tubers such as potatoes, however, may be successfully stored for periods extending to several months.

Although there are methods for preserving many of the perishable crops such as canning, freeze drying, etc., but these are normally industrialized processes and not found on farms. It is possible, however, to apply farm-scale methods of preservation to cereals and the less perishable root and stem tuber crops such as yam, cassava and potatoes. To do this successfully, it is necessary to know the ways in which a crop can deteriorate and hence the methods for controlling this deterioration.

Grains, hay and silage need to be properly stored to protect quality, reduce insect populations and maintain feed quality. Leak proof covers are essential for grain and silage, and are preferred for hay. Storage should not be in a location where water runs or ponds. Storage structures need to be secure and stable.

Good storage for farm crops is a sound investment, whether the material being stored is for use on the farm or for sale. Inadequate facilities and improper management can lead to substantial losses due to spoilage, insect and rodent damage, and fire from spontaneous combustion. Appropriate storage and handling methods should minimize losses, but must also be appropriate to other factors such as economies of scale, labour cost and availability, building costs, and machinery costs.

9.1 Temporary crop storage and adaptation

Storage in temporary structures is considered when permanent commercial and on-farm storage structures are filled. The primary problems in adapting existing farm shelters and machine sheds or using alternative structures for temporary grain storage include:

a. *Structural strength*; Stored grains in a building pushes out against the walls, so the building frame and lining materials must withstand heavy horizontal loads.

b. *Floor surface*; Existing buildings frequently have irregular and cracked concrete floors or earth floors. Without a vapour barrier, grain should be stored only 1 to 3 months on these floors. There may be some quality reduction due to spoilage and dirt contamination. A 4 to 6 millimeter plastic sheet can be installed as a vapour barrier on top of the floor surface.

c. *Grain handling method employed*; Filling and unloading of temporary storage are usually accomplished with a portable auger or other inclined conveyor. The auger can be easily shifted to shape the pile. Although fines distribution will be a problem, this will usually not be a big factor in 3months or less storage of cool, dry grain, unless the grain quality is very bad. Access to the grain should be planned for emptying. Alternative unloading tools include augers plus scoop shovels, a tractor with a front end loader, or pneumatic vacuum conveyor.

d. *Grain management*: Little aeration is needed for bins holding less than 26 m³ (1, 000 bushels). Larger bins require greater management, such as periodic monitoring and regulating of grain temperature and moisture conditions to ensure the quality of grain during storage. Excessive grain moisture usually causes moldy grain and provides conditions for insect attack that seriously decreases the value of stored grain. Important factors affecting grain quality during storage include the type of grain, beginning temperature and moisture, bin size, rate, time and method of grain ventilation, and length of storage.

Challenges of crop storage

Every storage structure, no matter what it looks like or what it is made of, should keep the product dry and cool, and protect it against insects, fungi, rodents, domestic animals and thieves.

1. *Temperature:* It is of greatest importance that a roof is constructed above the silos in such a way that it extends far enough to protect the walls against full sunlight. This lowers the inside temperature, lessens temperature changes between day and night and reduces the chance of local heating, which causes condensation in other, cooler spots with consequent fungus growth.

2. *Moisture:* Moisture may enter the storage container via the ground, the walls or the roof. A good overhanging roof protects the walls against rain. In order to prevent moisture coming up from the ground, the gourds, baskets, sacks etc. should always be placed on a dry underfloor or on a platform of bricks or wooden poles.

3. *Rodents:* Mice and rats should be combated by closing all holes in the silo or fitting these with fine-meshed wire netting, or by building the silo on poles, at least 75 cm

high, with rat baffles around the poles (Figure 9-2). The baffle must fit tight to keep even the smallest rodent from climbing between the baffle and the pole. If the silo is under a tree, rodents may jump from the tree and enter the silo via the roof.

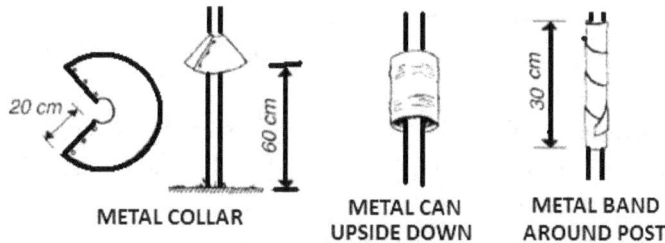

| METAL COLLAR | METAL CAN UPSIDE DOWN | METAL BAND AROUND POST |

Figure 9-2: Rat baffles

4. *Domestic animals:* These can be kept away by building a fence around the silo, made out of wood, bamboo or other local materials. A stack of twigs around the silo is also sufficient. Take care that it does not become a hiding place for rats.
5. *Insects:* First of all it is important to keep the storage room and the surrounding area as clean as possible, especially when using a non-airtight storage method. One should distinguish between insects that are already present in the product to be stored and insects that may enter the storage room during storage.
6. *Thieves:* Theft is made more difficult if the fill- and outflow openings are made in such a way that they can be locked. It is also possible to construct a silo without an outflow.

Requirement for good storage

Some of the requirements for a good storage system include:

    a. Protection from insects, rodents and birds by allowing proper storage hygiene.
    b. Ease of loading and unloading.
    c. Efficient use of space.
    d. Ease of maintenance and management.
    e. Prevention of moisture re-entering the grain after drying.

Planning grain storages

Consider the following in planning for grain storage

1. *Building site:* Surface and groundwater drainage is essential. Choose an elevated site or, if necessary fill the site with compacted gravel or sand before building. Arrange the buildings so that grain can be moved mechanically from one to

another or from any building to a vehicle. The storage area should have access to a public road.

2. *Floors and foundations*: Since grain is heavy, a reinforced concrete floor on the ground is recommended and is usually the most economical. The concrete floor should always be placed over a moisture barrier (such as a sheet of polyethylene) on compacted sand fill.

3. *Anchoring*: Anchor all bins securely so that no shifting is caused by wind when empty or by the grain when filled. Wood wall studs must also be securely fastened to the sill and plate.

4. *Grain handling:* Grain is easy to handle mechanically. Arrange bins systematically so that the grain can be moved from one bin to another by horizontal or inclined conveyors (such as augers), often combined with a bucket elevator for vertical lift. Mechanical handling is essential for moving grain from trucks to storage quickly and efficiently, particularly during harvesting and drying operations when time is critical.

5. *Care of dry stored grain:* Grain should be stored clean and dry, and inspected periodically for hot spots, dampness or other signs of moisture migration or spoilage. Storages over 100 m³ should have perforated floor ducts for periodic fan aeration, to redistribute or remove moisture that migrates within the grain mass.

6. *Rodents:* To control rodents, clean up weeds and debris that can provide hiding places around the grain center. Build all access openings rodent-tight (for example, use galvanized steel flashings, carefully fitted).

Methods of crop storage

Storage methods can be divided into airtight and non-airtight storage.

Airtight storage can be achieved using pots and gourds that are varnished or treated with linseed oil, pitch, bitumen or any thick substance which will stick to it.

Non-airtight storage is that storage which allows airflow through the products while in storage.

Objective of storage

The objective of storage is to maintain the quality of the crop during the storage period, either short-term (2–6 weeks) or long-term (over 4–8 weeks or more). To keep grain in good condition, it should be stored at a relatively low moisture content and cool temperature in order to prevent the development of molds and insects. Before building new storage facilities, a complete storage and handling system should be designed to meet both present and future requirements and to maximize mechanization.

Crops storage structures

Products can be stored in many different kinds of storage containers varying from earthen gourds, baskets, cribs and suchlike, to big metal or cement silos. Depending on financial possibilities, available materials and external circumstances (climate) one can choose from the storage methods mentioned above. The number and size of storage buildings/structures depend on the quantity and kinds of crop to be stored. Farm storage structures/methods include:

Traditional storage

For centuries people have used different methods to store grains - depending on local economy and climatic factors. Grain storage occurs in different stages– at field level when shoots are cut and dried before threshing and at field level when grains are kept or several days during threshing season before being taken to farmer's compounds. At every stage care is taken to protect grains against damage. Grain is affected by moisture level, especially at early stage of harvest, aeration process, sunlight, temperature, microorganisms and various diseases and pest attacks. Women play an important role in traditional grain storage everywhere and they are familiar with different stages and steps of storage process.

Traditional storage practices

There are several traditional or local methods identified for the storage of agricultural crops. These include:

Earthen pots and gourds

Especially in the dry tropics, earthenware pots and gourds (the hard, dried outside cases of certain fruits or vegetables) are very useful for storing small quantities. They should be kept inside or under a shelter. Above the kitchen (a place where there are few insects) is a good place. By treating the pots and gourds with varnish, paint or linseed oil, and sealing the lids with mud or cowdung an airtight form of storage is obtained (also suitable for the wet tropics).

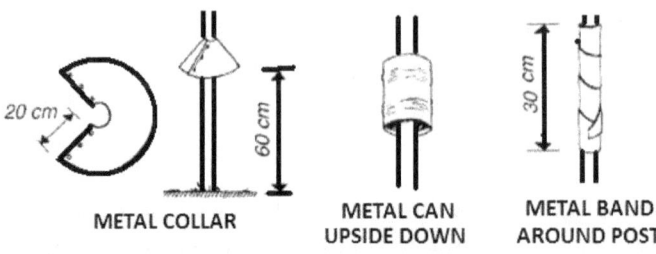

METAL COLLAR   METAL CAN UPSIDE DOWN   METAL BAND AROUND POST

Figure 9-3: Earthen pots

Mud/clay jars

A common method of building mud/clay jars is to prepare them a little above ground to avoid contact with the soil and thus preventing moisture and access by pests, insects and rodents. Moreover, the clearance from the earth allows aeration and cross ventilation in order to maintain a suitable temperature inside the jar.

Figure 9-4: Clay grain storage

To strengthen the walls of the clay jar mud is mixed with fine wheat straw and occasionally ropes and threads of palm or similar plant are used for further reinforcement. This storage structure is suitable for small quantities of cereals, beans, groundnuts, and also for the storage of sowing seed. The storage time is about one year

Baskets

Baskets are especially usable in the dry tropics. Baskets do not give enough protection against insects, but this can be improved by applying mud, clay or cowdung to the in- and outside. The cover should be tight, and sealed with plaster of the same material. The same effect is obtained by using a plastic bag inside the basket, which also makes storage airtight.

Figure 9-5: Traditional baskets

Improved traditional basket has plastered clay walls, inside and outside, raised off the ground on wooden platform standing on baffled poles and traditional grass or reed thatch roof. An emptying spout made of tin with press-on lid (e.g. coffee tin) with bottom removed is set into wall.

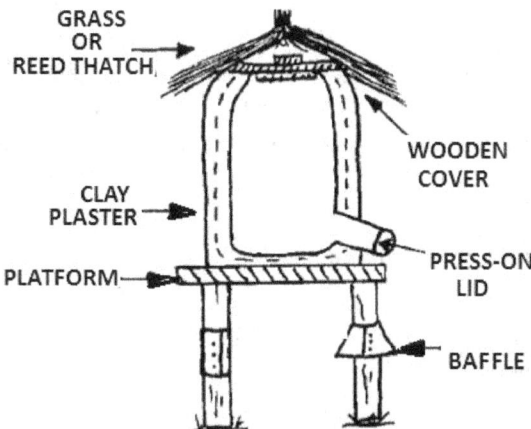

Figure 9-6: Improved traditional basket

Mats made from palm leaves

The second method is to store grains in a mat container made out of dwarf palm leaves or date palm leaves. Several mats are sewn together to form a round jar 1.5 meters in height and 5-6 meters in circumference. The size of jar depends upon the amount of grains and is made accordingly. The mats have the quality of providing aeration to some extent and are of organic materials which help to minimize the effects of high and low temperature. Care is also taken to keep domestic animals away from the storage. Before placing grains in the jar a straw layer of few inches is placed underneath and direct contact with bottom of the jar is avoided.

In some cases grains are stored in heap of wheat straw without an inside structure made from local available thatches and reeds. In such cases the heap of wheat straw is pressed on the ground and a bowl shaped depression is made in the centre of this heap and grains are kept inside and then covered by heavy layers of straw. This is basically for larger amounts of grains than household level. Farmers believe that in this way pest attack is minimized.

Underground storage for grains

Grains are also stored in the ground by digging a circular ditch. Special care is taken while selecting the site to avoid flooding, seepage or collapse of soil and quality of soil such as access of salts etc. to avoid any possible contact and reaction with grains. Usually this is practiced in more arid areas and desert region and is called *'khurrum'*.

In general special care is taken with respect to the timing of storage. Humidity in the air is avoided and sun light is preferred during the storage operation. Additional coverage is provided by cloths and plastic sheet during rains and storms. It is made sure that no drainage water drainage passes near the storage site. The storage site is detached all around. In areas of high velocity winds additional support of beams or stones are provided. Even storage facilities are affected by humidity and losses are common.

Traditional storage structures for tuber crops

The following storage structures have been used in storing tuber crops effectively in the topical environment

Underground pit storage

In parts of Africa, India and Latin America, underground pits are claimed to have keep grains/tuber crops without damage for many years. The pits keep grain cool, and some of them are relatively airtight. Grain on top and around the sides can however often be mouldy. There are several types of pits, most of them flask shaped covered with sticks, cow dung and mud, or a large stone embedded in soft mud. The area should be free from termites and relatively dry.

Burying harvested roots underground is a common storage method. Roots also are piled in heaps and watered daily to keep fresh. Several structures and methods have been utilized to improve the storage life and minimize spoilage of cassava roots, ranging from trench pit and boxes containing sawdust to bamboo-thatch room and permanent buildings. Weight loss and microbial losses are high in these structures due to poor environmental control.

Figure 9-7: Crop storage in trench pit

Roots in sawdust storage should be inspected every 3 days to ensure that sawdust is kept moist, while the top soil layer of the trenches should be moistened regularly once a week with clean water. "Damp storage" structures developed by IITA comprises of a pile of roots covered by straw and then soil, and ventilators then are used to reduce temperature in the pile during the hot dry season.

Traditional lining of pits is done with plant materials, such as grass, straw, chaff, maize stalks or sorghum and/or clay, cowdung or termite mound soil which is usually "fired" to harden the surface. The use of plant materials alone probably does little more than prevent the grain from coming into contact with the soil, unless applied very thickly. The use of clay etc. may also restrict water entry, but does not prevent it entirely. The pits are closed and sealed with plant materials and soil, or with stones, sticks or bark, usually plastered over with fresh cowdung. Pits may be cylindrical, rectangular or narrow- necked and may be placed in raised ground, under dwellings or beneath puddled clay or dung layers to prevent penetration of water.

Improved pit storage

The roofing of the pit can be made of metal sheeting, sealed with mud/dung or bitumen, or polythene sheeting. A temporary shelter over the pit site gives protection from rain but should be removed in the dry season to ensure drying by evaporation, because a shelter does not prevent lateral movement of water into the pit.

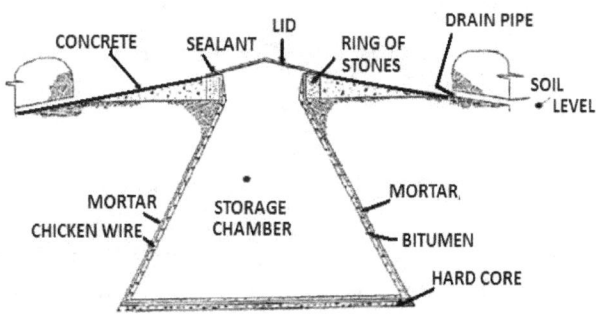

Figure 9-8: Improved ferro cement lined underground tip

Yam barn

The yam barn is the principal traditional yam storage structure in the major producing areas. Barns usually are located in shaded areas and constructed so as to facilitate adequate ventilation while protecting tubers from flooding and insect attack (Figure 9-9). Harvested yams can be put in ashes and covered with soil, with or without grass mulch as shelter. "Yam houses" have thatched roofs and wooden floors, and the walls sometimes are made simply out of bamboo.

Figure 9-9:

Traditional yam barns *(IITA flicker)*

Bag storage

The most common method for grain storage in many countries is bag storage in a variety of buildings, e.g. stone, local brick, corrugated iron, and mud and wattle, with or without plastered walls and with earth, stone, or cement floor and corrugated iron or thatched roof. However, if the grain is going to be kept for some time it is recommended to store the bags in a building. In general just sacks are cheaper than sacks made of cotton or sisal.

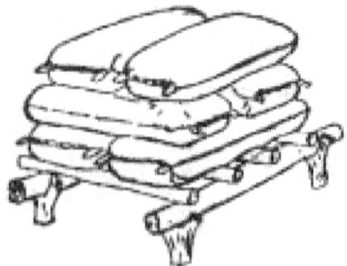

Figure 9-10: Pile of sacks on a pallet

These sacks are especially suitable for the dry tropics. Because of the danger of moisture uptake they should not be placed on concrete floors or on the ground, but on plastic sheets, waterproof canvas or on wooden pallets.

The latter method is preferred because it also allows air to flow under the sacks. Do not stack sacks against the walls, as insects and termites can get into the grain from the walls. Stack the sacks in a neat manner in not too big quantities on top of and against each other. Leave some space between the sacks so that air can move freely between them (Figure 9-11).

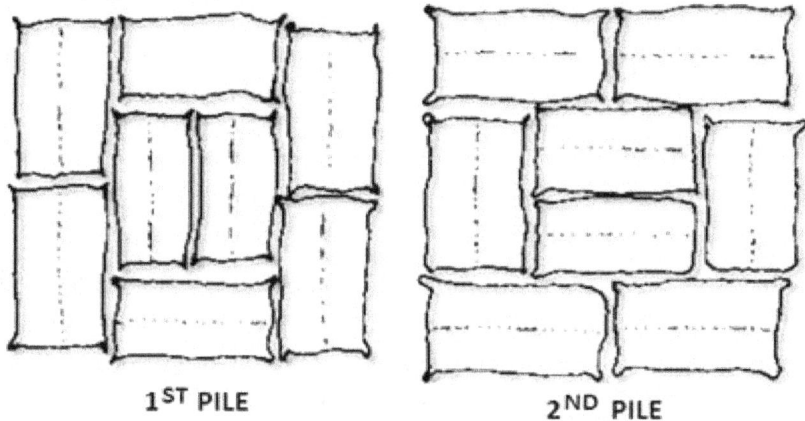

1ST PILE   **2ND PILE**

Figure 9-11: Stacking pattern

Advantages and disadvantages of fiber sacks

Advantages of fiber sacks are the following.

1. The product can have a slightly higher moisture content than when put into airtight storage, provided the sacks are stacked in such a way that air can move through the sacks for continued drying and cooling.
2. Sacks are easy to handle and label.
3. These sacks allow gasses to pass through and therefore insects may be controlled by using fumigants in a closed room or underneath a plastic sheet covering the stack.

The main disadvantage is that

1. Fiber sacks do not give much natural protection against insects, rodents, fungi and moisture.
2. They are also easily damaged during transport and handling.

In general terms the advantages and disadvantages of bag and bulk storage respectively, are as presented in the following table:

Table 9-1: Comparison between bag and bulk storage

Bags	Bulk
Flexibility of storage	Inflexible storage
Partly mechanizable	Mechanizable
Slow handling	Rapid handling
Considerable spillage	Little spillage
Low capital costs	High capital costs
High operating costs	Low operating costs
Easy inspection	Inspection more difficult

Crib storage

Cribs are enclosed structures (in different shapes) elevated between 0.5m and 1.0m above the natural ground elevation and supported on columns to guide against ground moisture, aid ventilation and protect the stored produce from insects and rodents. Locate ear-corn cribs in a north-south direction on a well-drained site with year-round accessibility. Place them at least 20 m (65 ft) from other farm buildings to ensure maximum wind effects and to minimize snow drifts.

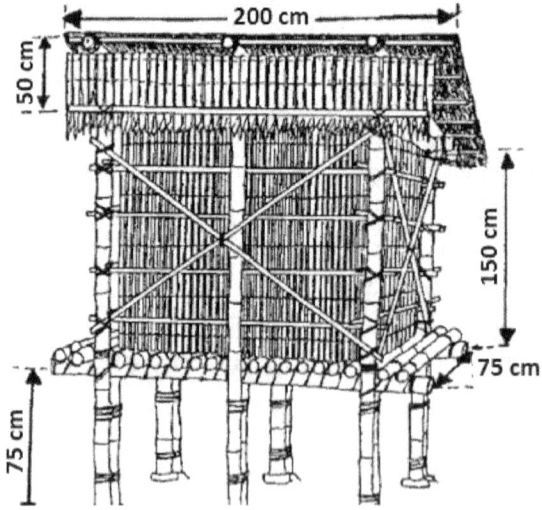

Figure 9-12: Ventilated maize crib made of bamboo (Wilson and Victor, 1980)

The maize crib is suitable for the storage of maize cobs in the humid and dry tropics. The maize crib design of Figure 9-13 has such a shape that the drying process continues during storage, because of natural ventilation. This applies to periods of low rainfall, during which the relative humidity regularly drops below 70%. A simple store would be to use the ventilated maize crib that was used for drying, with the only difference being that the walls should be covered as protection against rain (Figure 9-13).

Figure 9-13: Improved maize crib

The crib should be placed ideally with the long side perpendicular to the prevailing wind direction during storage. Because of the narrow shape (maximum width in the humid tropics is 60 cm) the drying process is better than in the traditional round crib. The maize crib offers very little protection against insects. The maize is not protected against thieves if it is stored in this way. It is suitable for storage of maize and yam and the storage time is between 3 - 6 months without insecticides.

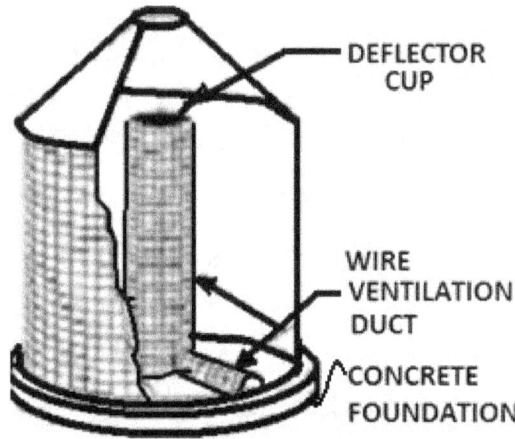

Figure 9-14: Metal crib

9.2   Storage bins

The traditional bins used by the African farmers are small with capacity of up to 2-3 tonnes and include gourds, clay pots, mud plastered baskets raised off the ground and mud walled silos ("rhumbus') (Figure 9-15).

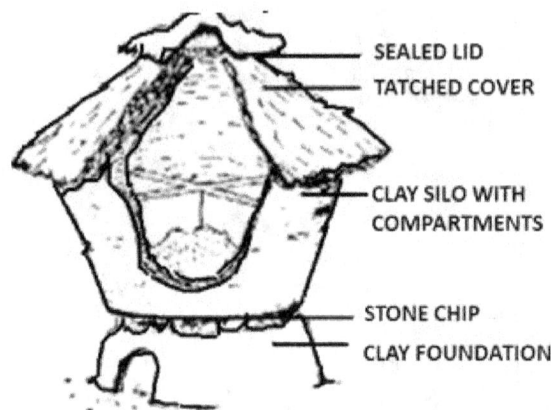

Figure 9-15: Clay silo (rhumbus) for storing grain

Many of these have limitations, particularly in durability, protection against rodents and insects as well as moisture from ambient air. Solid wall bins or silos should only be used in areas where the produce can be dried sufficiently before storage. Several attempts have been made to improve on the traditional stores to make them more suitable for long term storage.

Improved traditional bins

Much traditional storage performs excellently in their appropriate climatic conditions and others can be made to do so with minor changes. Efforts should be made to prevent cracks in the surface of the walls and to seal the entrance to the bin. This can be done for instance by adding lime or cement to the mud (i.e. a stabilized soil technique) or by incorporating an airtight lining (e.g. plastic) in the wall.

Figure 9-16: Improved traditional bin

Figure 9-16 shows a woven basket made of sticks or split bamboo plastered with mud mixed with cement. The walls slope towards a covered manhole in the top. An outlet is near the bottom the bin, which is placed on a raised platform, is covered by a thatch roof or hat.

Single-purpose circular grain storage

Concrete or steel silos are sometimes used for storing dry or high moisture grains. Concrete silos generally need to be reinforced and weather proofed to hold grain. A weather-tight chute and roof, along with foundation drainage are essential.

Grain flow analysis in storage bins

Designing systems for optimum grain flow in a storage bin starts in the field. Increases in acreage, harvesting capacity, and hauling equipment size are required in close analysis of the entire system. Bottlenecks are reduced by considering the followings:

1. *Receiving:* The material is received with an auger conveyor from the receptacle or temporary storage.
2. *Elevating or lifting:* Grain lifting is achieved with a bucket elevator system, which has required capacity. Top priority is to handle grain at the same rate as it is being received or loaded.
3. *Conveyance:* The products could be conveyed to storage bins with augers, U-troughs, or drag conveyors, to achieve the perfect fit between wet and dry handling systems.

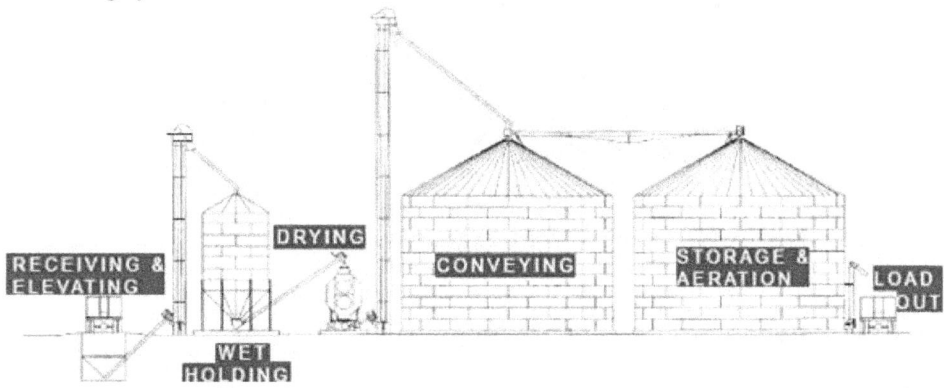

Figure 9-17: Grain flow in silo storage system

4. *Drying system:* External and in-bin continuous-flow systems are key components for expanding drying capacity to meet today's harvest rates and also automates the drying and handling operation. In-bin continuous flow drying systems adapt to new as well as existing drying bins. Capacities range from 100 to 500 BPH depending on bin size and fan/heater combination (6' corn depth, 25 to 15 full heats and cooling in storage.) Automated drying and automatic transfer to storage reduce operational and handling labour.

5. *Wet holding:* Harvested grains waiting to be moved through the dryer are at the heart of every grain system. Proper sizing keeps the dryer running and prevents waiting to unload.

6. *Storage, aeration and load out:* Properly sized storage combined with a well planned aeration system offers flexibility to meet marketing needs, reduce labour, and control grain quality.

9.3 Types of silos

Silo designs can be any of the following types,

a. *Horizontal silos*: These may be the trench or pit silo, made either by excavating earth below grade or by making an earth embankment, or the bunker silo, made by erecting concrete or timber retaining walls above grade.

b. *Tower silos*: Tower silos are built with cast-in-place concrete, concrete staves or steel. They may be exposed to outside air at the top (with top-unloader), or sealed to control the inside atmosphere (usually with bottom unloader).

Figure 9-18: Steel grain silo system

c. *Reinforced concrete silos*: Concrete can take very little tension and needs to be reinforced when used for silos. Small silos suitable for farm level may be reinforced with chicken-wire. The ferrocement store or ferrumbu (Figure 9-19) is a typical example.

Figure 9-19: Cross-section of a ferro-cement store

Larger concrete silos are built by using a sliding mould which is moved upwards continuously or step by step. Reinforcement and concrete are supplied from the top concrete silos can be made airtight if openings are properly sealed.

9.4   Rectangular multipurpose buildings

These are large volume storage facilities popular among large scale farmers, cooperative storage centers, local state and federal government (e.g. strategic grain reserves). Warehouses vary in capacity and can be up to 5000 tonnes.

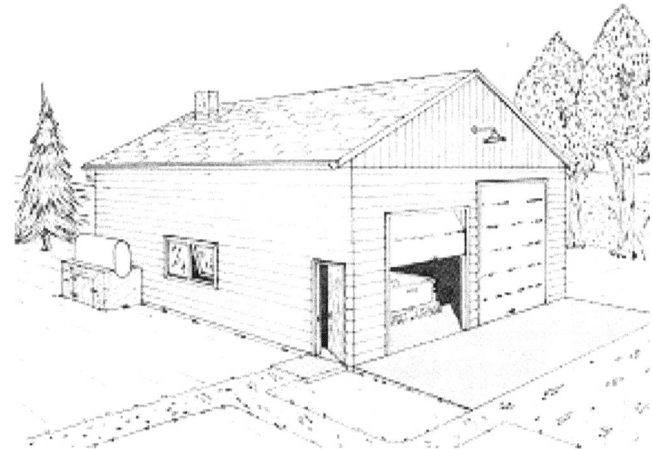

Figure 9-20: Farm warehouse

9.5   Hay and straw storage

 Hay and silage need to be properly stored to protect water quality, reduce fly populations and maintain feed quality. To reduce spoilage and weathering of hay and straw, provide some form of shelter. Simple pole-type construction with a plywood or metal roof is economical and reasonably effective. Select a well-drained site to prevent water accumulation at the base of the feed piles. Leak proof covers are essential for grain and silage, and are preferred for hay. Storage should not be in a location where water runs or ponds.

Further reading

Bello R. S., 2007. *Fundamental Principles of Agricultural Engineering Practice.* Enugu Nigeria Climax Printers,

Bello R. S., 2012. *Agricultural Machinery & Mechanization*. 7290 B. Investment Drive Charl 7290 B. Investment Drive Charl createspace ISBN-13: 978-1456328764. https://www.createspace.com/3497673

CIGR, 1999. CIGR Handbook of Agricultural Engineering. Volume IV Agro-Processing Engineering. Edited by CIGR–The International Commission of Agricultural Engineering. Published by the American Society of Agricultural Engineers

CHAPTER 10

Animal Housing and Requirements

10.   Introduction

The main purpose of man keeping livestock is to convert energy in the feed into products which can be utilized by human beings, such as milk, eggs, meat, wool, hair, hides and skins, draught power and manure (fertilizer). Traditionally, extensive livestock production involving indigenous breeds and low cost feeding usually have low performance and can therefore only justify minimal, if any, expenditure for housing.

However, where improved breeds, management and feeding is available it will usually be economically beneficial to increase the production intensity and to construct buildings and other livestock structures to provide for some environmental control, reduced waste of purchased feed stuffs and better control of diseases and parasites.

10.1   Animal behaviour and building design

Animals that can exercise their natural species-specific movements and behaviour patterns are less likely to be stressed or injured and will therefore produce better. In practical design of an animal production system and any buildings involved, many other factors such as feeding, management, thermal environment, construction and economics can be equally more important.

The importance of animal behaviour aspects in the design of animal housing facilities generally increase with the intensity of production and the degree of confinement. Many modern farming systems greatly reduce the freedom for animals to choose an environment in which they feel comfortable. Instead they are forced to resort to an environment created by man which is against the growing campaign against livestock captivity, since they believe naturally that animals enjoy relative freedom in the wild before domestication.

10.2   Animal thermal environment

An animal thermal environment is the total of all external environmental conditions that affect an animal and its environment. It concerns the heat transfer, the climatic factors of the air, temperature moisture, air velocity and solar radiation that affects the regulation and balance of animal heat and those that influence production growth, feed conversion and health. The total of heat production in animal environment is expressed as:

$$Heat\ Production\ =\ Heat\ Loss \pm Heat\ Storage \dots\dots\dots\dots\dots\dots..10.1$$

The rate of internal heat production varies with size, body weight, breed, health, state of growth, nature and rate of feed intake, level of production, gestation, age, degree of activity and environmental condition. 25-40% of total feed intake by an animal is converted to heat and loss to the environment.

Air exchange in livestock building

Air exchange is the portion of the livestock housing system that can be manipulated, which need to be controlled. The heat input and output for a livestock or poultry building must be equal under steady state environmental condition. A predicable amount of sensible heat will be transferred through the walls and exposed surfaces of the building. The remaining output of heat may be accounted for through air exchange. The amount of conduction heat loss through the exposed surfaces of insulation, shape of the building and the side temperature should also be accounted for. It should be noted size and width of buildings are not very significant from the standpoint of heat loss.

Ventilation

Ventilation is defined as a system of air exchange, which accomplishes one or more of the followings.

1.   Provides desired amounts of fresh air, without to all parts of the shelter.
2.   Maintains temperatures in the shelter within desired limits.
3.   Maintains relative humidity in the shelter within desired limits.

All ventilation systems once were natural but, with the availability of electricity on farms and a sensor control technology, mechanical or fan-powered ventilation took over. This shift from natural to mechanical ventilation took place mainly because building designers felt that the performance of mechanical systems was predictable, controllable, dependable, and effective and that natural ventilation was none of these.

Natural ventilation

Natural Ventilation of buildings is generated through two distinct sources;

a. Buoyancy or gravity effects are present, and in large part these are due to temperature differences between the outside and the inside air.
b. The second source is wind blowing over a building, generating pressures and suctions at different points, which can force air in and out of the building.

Forced ventilation

Forced ventilation means that the air-exchange rate in ventilated rooms is established by means of fans. It differs from natural ventilation in that it is less dependent on wind velocity and buoyancy and in the continued energy consumption.

A ventilation system always includes two main parts: inlet and outlet. In the case of forced ventilation at least one of them contains fans. If the fan is placed in the outlet system, it is called a *negative-pressure system*. If the fan is placed in the inlet, it is called a *positive-pressure system*, and if there are fans in both inlet and outlet it is called an *equal-pressure system*. The three principles are illustrated in Figure 10-2 below: (Left to right: a) negative-pressure system, b) positive-pressure system, c) equal-pressure system)

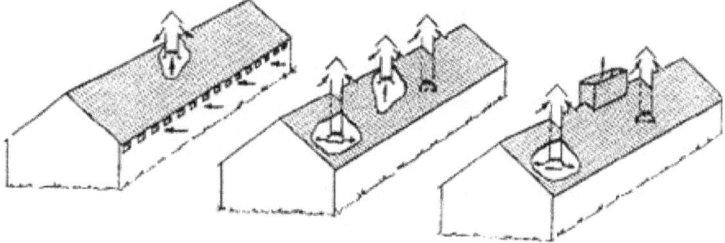

Figure 10-2: Ventilation principles

Building requirement

In the design of animal house, the following requirements are particularly essential and must be put into consideration:

1. *Heat transmission through building materials*: The rate of heat transmission through building materials depends on the characteristics of the material. A comprehensive

list of heat transfer coefficients of building material is given in the ASHRAE handbook of fundamentals (1997).

2. *Vapour barriers*: Animal shelters of the totally enclosed and ventilated type may be subjected to high moisture conditions. Water vapour transfer is not a problem as long as the saturated vapour pressure within the enclosure components remains above the actual vapour pressure gradient. Moisture condensation is not desirable because it causes decomposition of building components and reduces the insulating capacity of much insulation. To keep insulations dry a vapour barrier should be applied on or near the side of the insulation with the highest vapour pressure.

3. *Condensation*: Estimating condensation on the inside wall surface may be estimated by using the condensation predication graph.

4. *Exposure factor*: this is the conductive heat loss from all animal shelter per animal or bird per degree temperature difference exposure factor is determined from the size of the building, the insulation value of the building and the number of animals housed.

The normal ranges of exposure factors for some animals are indicated in the table below:

Table 10-1: Exposure factor of some animals

Animal	Rectal temperature	Exposure factor range (w/m²°c)
Dairy cow	101.5°f	5.3 to 26
Beef cow	101.0°f	5.3 to 26
Pig	102.5°f	0.5 to 26
Sheep	102.3°f	-
Chicken	107.1°f	0.1to 26

The exposure factor can be expressed mathematically as

$$E_f = \frac{(A_i \times U_i)}{N} \quad \text{............... 10.2}$$

Where

E_f = Exposure factor

U_i = Overall coefficient of heat transfer for a particular structural component (i.e. walls, ceiling, doors, windows, etc). (W/m²°c)

A_i = Area of the specific building component for which the U_i value was determined. m²

N = Number of animal units housed (an animal unit for cattle is 1000 lb (45kg) animal and for swine and poultry it is an individual animal).

The total conducive heat loss from the shelter can be obtained by multiplying the exposure factor by the member of animals housed and the difference between inside an outside design temperature.

10.3 Livestock housing

Animals that have a variety of needs were made to live together in buildings which must satisfy those needs. The quality, number, and the characteristics of the individual buildings that form the overall system are necessary in order to identify the needs of each group of animals and the number of animals that it may contain.

The design procedure for defining the contents and size of the livestock buildings is dependent on the fundamental spatial elements that constitute variants of each production type for each different species of livestock. Input for the definition of the spatial elements involved in the production process is linked with the elementary activities of person, animal, machine involved.

Below are some examples of how animal behaviour that can influence the design of structures.

Cattle housing

Cattle normally live in herds, but when giving birth, the cow attempts to find a quiet, sheltered place away from the disturbance of other cows and humans. The cow needs to be alone with her calf for some time after birth for the cow-calf bond to be established. A cow approaching calving and confined in a loose housing system, should be removed from the herd and put in an individual pen.

Pig housing

Pig farming is relatively increasing in many tropical countries of Africa as processed pork finds an increasing market and pig production yields a relatively rapid rate of return on the capital employed. Pigs are kept primarily for meat production, but the by-products, such as pigskin, bristles and manure are also of economic importance. To some extent pigs compete with man for food, but they can also utilize by-products and wastes from human feeding.

Sows are nest builders and should be transferred to clean farrowing pens one to two weeks before giving birth, and given some bedding so that they can build a nest. Oestrus, especially in gilts, is increased by the smell, sight and physical presence of a boar. Gilts and sows ready to mate should therefore be kept in pens adjoining the boar pen.

To prevent waste of feed a trough should be designed to suit the particular behaviour pattern each species exhibits while feeding i.e. pecking in hens, rooting with a forward and upward thrust in pigs, wrapping their tongue around the feed (grass) and jerking their head forward in cattle.

Management practices

In many tropical countries pigs roam freely as scavengers or are raised in the back-yard where they depend on wastes for feed. Little attempt is made to obtain maximum In many tropical countries pigs roam freely as scavengers or are raised in the back-yard where they depend on wastes for feed. Little attempt is made to obtain maximum productivity. However, a few simple management practices can help to improve the productivity and health of these pigs. They include:

1. Fence paddocks with shade and water provided for the animals
2. Provide simple, semi-covered pens constructed of rough timber with thatched roof and concrete floors.
3. Wallows or sprinklers can be provided to alleviate heat stress. Being unable to sweat sufficiently pigs has a natural instinct to wallow to increase the evaporative cooling from the skin.

Figure 10-5: A pig pen

The raising of pigs in confinement is gradually replacing the old methods because of lower production costs, improved feed efficiency and better control of disease and parasites. Thus, the confinement system is usually advisable in circumstances where:

1. Good management is available;
2. High-quality pigs ate introduced;
3. Farrowing occur at regular intervals throughout the year;
4. Land is scarce or not accessible all the year;
5. Balanced rations are available;
6. Labour is expensive;
7. Parasite and disease control is necessary;
8. The target is commercial production;
9. Herd size is reasonably large.

Some systems keep only part of the herd in confinement. The order of priority for confinement housing for the different classes of animals is usually as follows:

1. Growing/finishing pigs (25-90 kg or more live weight) for higher control daily gain, better feed conversions and parasite control.
2. Farrowing and lactating sows, to reduce pre-weaning mortality and for higher quality weaners.
3. Gestating sows, to allow individual feeding and better control of stock.

General requirements for pig housing

A good location for a pig unit meets the following requirements: easy access to a good all-weather road; well drained ground; and sufficient distance from residential areas to avoid creating a nuisance from odour and flies.

An east-west orientation is usually preferable to minimize exposure to the sun. Breezes across the building in summer weather are highly desirable. A prevailing wind during hot weather can sometimes justify a slight deviation from the east-west orientation. Ground cover, such as bushes and grass, can reduce reflected heat considerably, and the building should be located where it can most benefit from surrounding vegetation.

A fairly light well drained soil is preferable, and usually the highest part of the site should be selected for construction.

Sheep and goat housing

Sheep and goats are important sources of milk and meat. Both readily adapt to a wide range of climates and available feed supplies. They also have similar housing requirements and will therefore be treated together. Housing in tropical and semi-tropical regions should be kept to a minimum except for the more intensive systems of production (Figure 10-6). In the arid tropics no protection other than natural shade is required. In humid climates a simple thatched shelter will provide shade and protection from excessive rain. Sheep and goats do not tolerate mud well; therefore yards and shelters should be built only on well drained ground.

Figure 10-6: Goat pens inside the house (FCAI)

Sheep are vigilant and tight flocking, and respond to disturbance by fleeing. When designing handling facilities these characteristics should be taken into account. A race should be straight, level, fairly wide, without blind ends, and preferably have close-boarded sides. Sheep which are following should be able to see moving sheep ahead, but advancing sheep should not see the sheep behind as they will tend to stop and turn around. Sheep move best from dark into light areas and dislike reflections abrupt changes in light contrast and light shining through slats, grates or holes. Handling facilities should be examined from the height of the sheep's eye level rather than the human to detect flaws in the design.

Rabbit housing

There are few, if any, countries where domestic rabbits are not kept for meat and pelts. It is widely recognized that a few rabbits can be kept for a low cost, but yet produce a fair quantity of wholesome and tasty meat. However, to raise rabbits successfully one must begin with healthy animals, provide a good hutch, clean and nutritious feed and take good care of the rabbits. There are certain essential features that any well designed hutch should provide:

1   Enough space for the size of the rabbit,
2   Fresh air and light, but exclusion of direct rays of the sun,
3   Protection from wind and rain,
4   Sanitary conditions and ease of cleaning,

5 Sound but cheap construction; which is free of details that could injure the animals,

6 Convenience of handling,

7 A cage for each adult rabbit.

Space requirement

Each adult rabbit must have its own cage or compartment. Since domestic rabbits vary in weight from 2 to 7 kilos, depending on breed, the size of cage may be determined by allowing 1200 to 1500cm^2 of clear floor space per kilo of adult weight. This means that a cage for a medium breed buck should be minimum 80cm square. However, cages for females should allow extra space for the nest box and the litter; hence 80 by 115cm should be regarded as minimum for a medium breed doe.

Figure 10-7: A rabbit hutch

Young rabbits reared for meat can be kept in groups of up to 20 to 30 animals until they reach four months of age. The weaned young kept in one group should be about the same age and weight. Such colony pens should allow 900 to 1200cm^2 floor spaces per kilo of live weight.

Figure 10-8: A rabbit

The cages should not be deeper than 70 to 80cm for ease of reaching a rabbit at the back of the cage. The floor to ceiling height of the cages should be minimum 45 to 60cm and it is desirable to have the floor of the cages 80 to 100cm off the ground to handle the rabbits comfortably.

Poultry housing requirement

Poultry, including chickens, turkeys, ducks and geese, offers one of the best sources of animal protein, both meat and eggs, at a cost most people can afford. Changes in poultry housing in recent years have been rapid and dramatic. The transition from the old farm chicken coop with a few hundred birds to a modern, environmentally controlled cage house for thousands of birds represent one of the greatest advancement ever made in housing for an agricultural enterprise. Automated equipment for feeding, watering, egg pickup, ventilating and manure removal has promoted egg production to one of the most efficient of farm operations.

Chickens are the most widely raised and are suitable even for the small holder who keeps a few birds that largely forage for themselves and require minimum protection at night. At the other extreme, commercial farms may have highly mechanized systems housing thousands of birds supplying eggs and meat to the city market. In between are farm operations in a wide range of sizes with varying types of housing and management systems proportionate to the available level of investment and supply of skilled labour. No single system of housing is best for all circumstances or even for one situation. Some compromise will invariably have to be made.

Figure 10-9: Poultry house

General housing requirements for chickens

Proper planning of housing facilities for a flock of laying hens requires knowledge of management and environmental needs in the various stages of the life of the chicken. Buildings for all phases of poultry production tend to produce considerable odour, hence, the site should be well downwind from living quarters. The best site is one that is well-drained, elevated but fairly level, and has an adequate supply of drinking water nearby (Figure 10-9).

Figure 10-10: Broilers in deep litter system (FCAI farm,)

Temperature is the most important environmental factors in poultry housing. Young chicks need very warm surroundings to survive. Older chickens, both layers and broilers exhibit their best feed conversion efficiencies at (21–24) °C; however, production drops rapidly as temperature rise above 27 °C and temperatures above 38 °C may be lethal. Humidity is important in two circumstances very low humidity tends to cause objectionably dusty conditions and high humidity combined with a very high temperature interferes with the birds' natural cooling mechanism and contributes to high mortality.

Regardless of the type or size of the housing system, the site for construction should be selected to provide adequate ventilation, but be protected from strong winds. An area under cultivation, producing low growing crops, will be slightly cooler than an area of bare ground. High trees can provide shade while at the same time actually increasing ground level breezes. Bushes planted at one windward corner and also at the diagonally opposite corner will reduce air currents within the building to make existing house shielded from the heat from direct solar radiation.

Hen brooding housing

Hens spend considerable time in the selection of a nest, which is on the ground. Nesting is characterized by secrecy and careful concealment. Hens in deep litter systems therefore, sometimes lay eggs on the floor instead of in the nest boxes, especially if the litter is quite deep or there are dark corners in the pen.

To avoid this, plenty of fresh litter is provided in the nests, and they are kept in semi-darkness and designed with a rail in front so that birds can inspect the nests prior to entry. An additional measure is to start with the nest boxes on the floor and slowly raises them to the desired level over a period of days.

Duck housing

Housing for ducks can be very simple. The house should be situated on a well drained, preferably elevated area. The floor should be raised at least 15cm above the surrounding ground level to help keep it dry. Ducks tend to be dirty and plenty of clean litter must be used in floor type housing.

Although a concrete floor can be installed for easy cleaning, it is not necessary. If part of the floor is of wire mesh and the ducks have to cross it on their way to the nest boxes, their feet will be cleaned so they do not make the nests and eggs dirty.

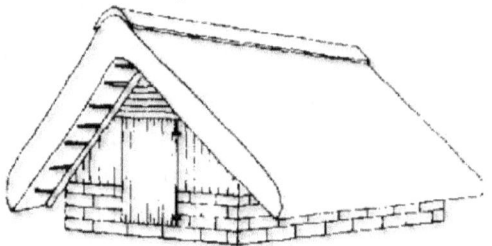

Figure 10-11: Duck house

Solid walls 60cm high are adequate. They may be made from any material as long as it keeps the ducks in and predators, like dogs, snakes, rats and wild birds out. The space between the wall and the-roof is covered with wire netting not larger than 25mm mesh. Total wall height does not need to exceed 150-200cm. Roof made of thatch is a fully adequate and inexpensive roof covering for a duck house. Metal sheets can also be used, but insulation should be installed under the sheets.

Nest boxes 30 cm wide, 40 cm deep, and 30 cm high should be provided for each four ducks. The front should be 15 cm high. The nest boxes are placed either on the floor or 30 cm off the floor against a side or rear wall. Although nesting boxes off the floor release more floor space, the ducks may lay their eggs under the boxes. Run and Fencing should provide a minimum of 1 m² per bird, but 2 to 3 m² or more, will keep the ducks cleaner and give more space for grazing. On open range pasture the ducks should be allowed 20 m²/bird.

Turkeys housing

Brooding and rearing methods for turkey poults (a young fowl, especially a turkey) are similar to those for chickens, but the brooding temperature is higher. The recommended temperature for the first week is 35° to 38°C, after which it can be reduced 4°C per week until ambient temperature is reached.

Adequate floor space in the brooder house is important as the turkey poults grow rapidly. At about 10 weeks of age, turkeys are put out on range in a fenced enclosure. In the interest of disease control, it is essential to use clean land that has not carried poultry, turkeys, sheep or pigs for at least two years. Approximately 20m² of pasture should be allowed for each bird.

A range shelter with 20m² of floor area is suitable for 100 poults up to marketing age. Dry, compact soil is adequate for a floor. The frame should be made of light material covered with wire mesh so that the shelter can be moved to clean range each year. The roof, which should be watertight, can be made of thatch or metal sheets. Perches, made from rails 5 x 5cm or round rails 5cm in diameter, should be installed 60cm from the ground and 60cm apart allowing 30 to 40cm of length per bird. The turkey breeder flock can be confined in a deep litter house similar to those for chickens.

10.4   Animal handling and examination

All species of livestock kept by farmer on the farm or reared for the benefit of man is threatened in some ways by disease and infections. Ticks have continued to be one of the most harmful livestock pests in Africa (Hall, 1986). As vectors of animal diseases, ticks have been a great hindrance to livestock development especially in areas where breeds of cattle exotic to the environment has been introduced. Presently, the only effective method of control for most of these diseases is control of the vector ticks. Dipping or spraying with drug remains the most efficient way of reducing the number of ticks.

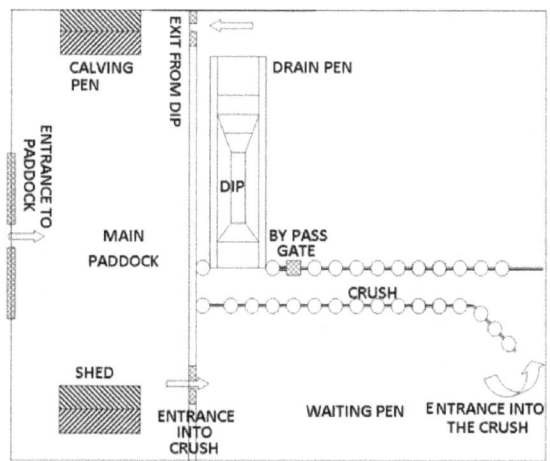

Figure 10-12: Schematic layout of a paddock

Handling cattle involves risk of injury from crushing, kicking, butting or goring. The risk is increased if it involves animals that are not being handled frequently, such as those from other farms, sucklers or newly calved cattle. Certain jobs, such as veterinary work, may increase the risk further. However, proper handling systems, training and competent staff, and a rigorous culling policy can help ensure that cattle handling can be carried out in relative safety (Bello, 2007, 2011).

The following methods of restriction for animal restriction and treatment were discussed:

Method of restriction by catching (R by C)

Traditionally, animals were restricted by getting people to pursue them with either rope or mere bare hands. To restrict animals in this way, an experienced catcher and the herder will pose with a knotted rope displayed in such a way that the animal's leg is restrained when the rope is thrown and thus overpowered. 'Restriction by catching' (R by C) is carried out by pursuing the animals within an enclosure (paddock) and restraining for examination. Handling an entire flock this way, especially when the animals are not fed before the exercise, they become aggressive and violent. It also takes up to 20-30 minutes to restrain one cow.

There have been increased concerns in some of the methods used in handling livestock, especially when large size of animals were involved over large expanse of land area. These concerns arose from a substantial number of reasons relative to efficient food production and handling systems adopted in their management which often cause stress in animals, loss in weight, abortion in pregnant animals, injuries inflicted on animals and 'catcher' and some other adverse physiological and psychological problems. Poor drug administration and drug waste in hand bathing caused reduction in effective drug management up to 20%.

Hand spraying

This is a method of treatment that can be effective if carried out by an experienced person on an animal properly secured in a crush. The cost of the necessary equipment is low, but the consumption of liquid and chemical is high as it is not re-circulated. The method is time consuming and therefore only practicable for small herds where there is no communal dip tank or spray race.

Figure 10-13: Hand spraying

Cattle dip

Dip is a structure built for the treatment of ectoparasites in brute animals like cow or other beast of burdens. Ticks continue to be one of the most harmful livestock pests in Africa. As vectors of animal diseases, ticks have been a great hindrance to livestock development especially in areas where breeds of exotic cattle have been introduced to the environment. At present the only effective method of control for most of these diseases is control of the vector ticks is dipping. Dipping or spraying with an acaricide is the most efficient way of reducing the number of ticks.

By a way of caution, cattle must not be allowed to get hot or thirsty when they are being dipped to prevent their gulping the treated water in the through, so it is important to have a water trough inside the collecting yard fence (paddock).

Dip design features

The following design features are common to all dip configurations

Footbaths: Footbaths are provided to wash mud off the feet of the cattle to help keep the dip clean. At least two are recommended, each 4.5 meters long and 25 to 30cm deep, but in muddy areas it is desirable to have more. Up to 30 meters total length may sometimes be required (Figure 10-13).

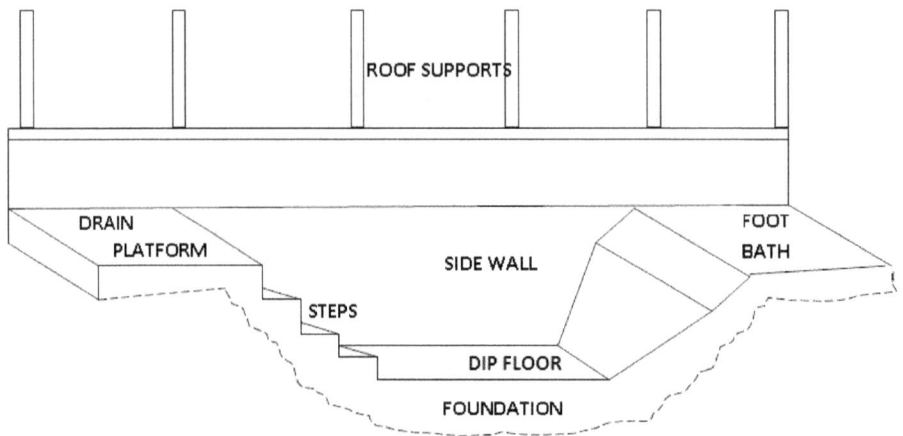

Figure 10-14: Cross section configuration of dip

The footbaths should be arranged in a cascade, so that clean water added continuously at the end near the dip, overflows from each bath into the one before it, with an overflow outlet to the side near the collecting pen. If water supply is extremely limited, footbath water can be collected in settling tanks and reused later.

The floor of the baths should be studded with hard stones set into the concrete to provide grip, and to splay the hoofs apart to loosen any mud between them. The side walls to the floor slopes backward to form a trapezoidal shape with the floor. The top width is 1.25m (4.1ft).

Jumping place: This is a narrow steep flight of short steps ensures:

- That animals can grip and jump centrally into the dip,
- That their heads are lower than their rumps at take-off,
- That they jump one at a time, and
- That dip-wash splashing backwards returns to the dip.

The jumping steps are approximately 0.4m wide and 0.15m rise out of the dip into the drain platform. The jumping place of 0.4m above the dip-wash level is desirable to give maximum immersion, there could be some danger to heavily pregnant cows if the water level was allowed to fall a further 40cm. (The dipping of 1,000 cattle without replenishment would lower the water level to 60cm below the jumping place). The lip of the jumping place experience extreme wear and should be reinforced with a length of 10cm diameter steel pip.

Splash walls and ceiling: A splash wall and ceiling are provided to catch the splash and prevent the loss of any chemical. The ceiling will protect a galvanized roof from corrosion.

Roof: In a typical dip, the roof is 1.83m (6ft) above the dip and a .76m (2.5ft) ridge spanning the entire length of the dip 10.36m (34ft) (Segun, 2011). The ridge slopes down at both sides of the roof made of corrugated iron sheet. The walls can be made of wood, but masonry is most durable.

The dipping tank: The dipping tank is designed to a size and shape to fit a jumping cow and allow her to climb out, while economizing as far as possible on the cost of construction and the recurrent cost of chemical for refilling. A longer tank is needed if an operator standing on the side is to have a good chance of re-immersing the heads of the animals while they are swimming, and increased volume can slightly prolong the time until the dip must be cleaned out.

Poured reinforced concrete is the best material to use in constructing a dipping tank in any type of soil although expensive if only a single tank is to be built, because of the cost of the form-work involved, the forms can be reused. If 5 tanks are built with one set of forms the cost per tank is less than the cost of building with other materials, such as concrete blocks or bricks.

Figure 10-15: Cross section configuration of dip

A reinforced concrete dipping tank is the only type with a good chance of surviving without cracking in unstable ground. In areas prone to earthquakes a one-piece tank is essential. All dipping tanks need to be cleaned out from time to time and disposed of the accumulated sediment. It is normal for all the waste dip-wash to be thrown into a 'waste pit' that is dug close to the dip. In addition dipping tanks may crack with leakage of chemical as a result.

Trough: Cattle prefer to be able to see while drinking; therefore more animals can drink at once from a long, narrow trough than from a low round one. With cattle (and hens) feeding is typically a group activity; therefore space at the feed trough must be provided for all the animals at one time.

Catwalk and hand rails: These are provided to allow a person to walk between the splash-walls to rescue an animal in difficulty. In addition to providing shade, a roof over the dipping tank reduces evaporation of the dip-wash, prevents dilution of the dip-wash by rain, and in many cases, collects rain water for storage in a tank for subsequent use in the dip.

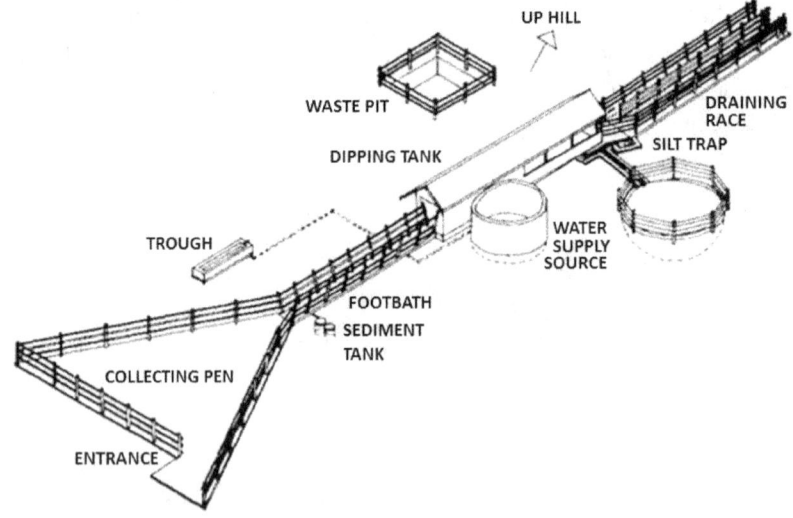

Figure 10-16: Cattle dip layout. (Adapted from Lennart and Whitaker, 1998)

Draining race: The return of surplus dip-wash to the dipping tank depends on a smooth, watertight, sloping floor in the draining race. A double race reduces the length and is slightly cheaper in materials, but a very long single race is preferable where large numbers of cattle are being dipped. Side-sloping of the standing area towards a channel or gutter increases the back-flow rate. The total standing area of the draining race is the factor that limits the number of cattle that can be dipped per hour, and the size shown in the drawings should be taken as the minimum.

Collecting pen: This is a temporary stop over place for the cattle before entering the footbath. The floor is constructed with concrete. The shape and size requirement depends on choice and available land area.

Silt trap: A silt trap allows settling of some of the mud and dung from the dip-wash flowing back to the tank from the draining race. The inlet and outlet should be arranged so that there is no direct cross-flow. Provision must be made to divert rain water away from the dip.

Land area (site) requirement: Minimum land area requirement for dip construction is 30 x 80m (1 acre). *Road* accessibility for vehicles and animal must be adequate and unobstructed.

Water requirements: Annual consumption per animal per week should be estimated as well as dipping tank water requirement must be estimated and adequately provided for.

Water supply sources: Requirement for sources of water supply to the paddock could be through one or more of the followings:

a. Roof catchment through rainfall
b. Water collected from draining race, roof or road catchment tank
c. Water supply from shallow well and
d. Water from gravity dam, siphon, hand pump or water supply schemes tec.

Cattle crush

A crush can vary in length, but should not be shorter than two cow lengths because cattle have a herding instinct, and should an animal refuse to enter a crush, placing a tame animal in front often entices other animals to follow.

Figure 10-17: Typical crush

Where a crush is more than 14 animals long, efficiency is lost because people working with the cattle need to walk too far from the pre-holding area to the front of the crush and, unless spacers are provided, cattle in a long crush tend to move back when handled and crush animals in the rear. A crush length of 1 700 mm per medium sized cow is usually satisfactory. Vertical posts should be placed at one animal interval.

Many farmers like to provide for a smaller crush where calves can be handled. However, a full-sized crush can be used for calf management by adding barriers lower down. Stringing a cable between the ground and lowest horizontal pole and between the first and second horizontals prevents calves from escaping. The handler then enters the crush and works with the calves in the crush, with the neck clamp closed and the rear of the crush closed with poles. With larger calves, crush width can be reduced by hanging motor-car tyres or poles inside the crush.

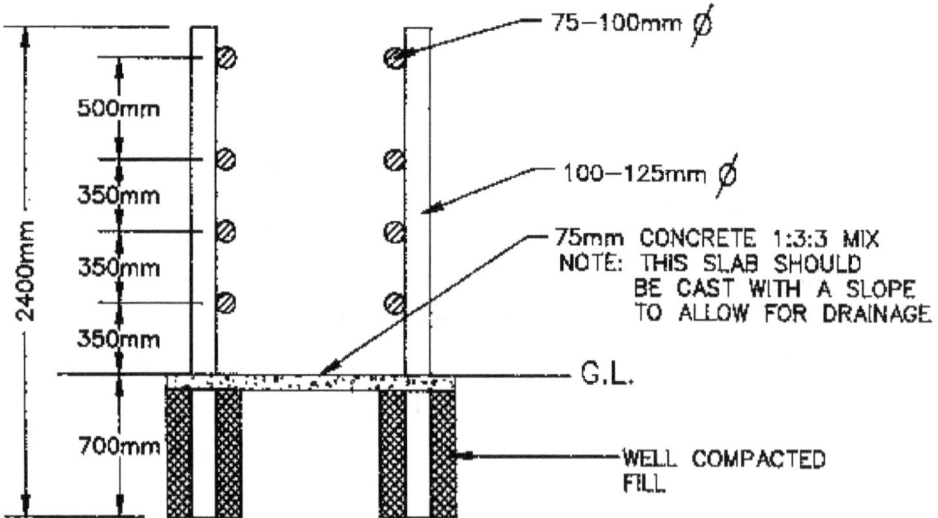

Figure 10-18: Sample spacing of crush

Although a neck clamp is useful, a crush with a gate in front can be used for most operations by using some means to immobilise animals, such as tying their heads to the side of a crush using a leather thong, which will not cut into their skins. Loose poles placed across the crush in front of vertical posts are usually all that is needed to prevent animals exiting at the rear. Figure 9-18 illustrates the spacing of horizontals and verticals in a crush. Verticals must be firmly bedded, preferably in concrete.

Cattle spray race

A spray race site requires the same features as a dip site and these have already been described. The only difference is that the dip tank has been changed for a spray race. The race consists of an approximately 6m long and 1m wide tunnel with masonry side walls and a concrete floor. A spray pipe system on a length of 3 to 3.5m in the tunnel having 25 to 30 nozzles installed on the walls, ceiling and floor, will discharge dip liquid at high pressure and expose the cattle passing through to an intense spray. The fluid is circulated by a centrifugal pump giving rise to continuous flow and recirculation of drained or used water.

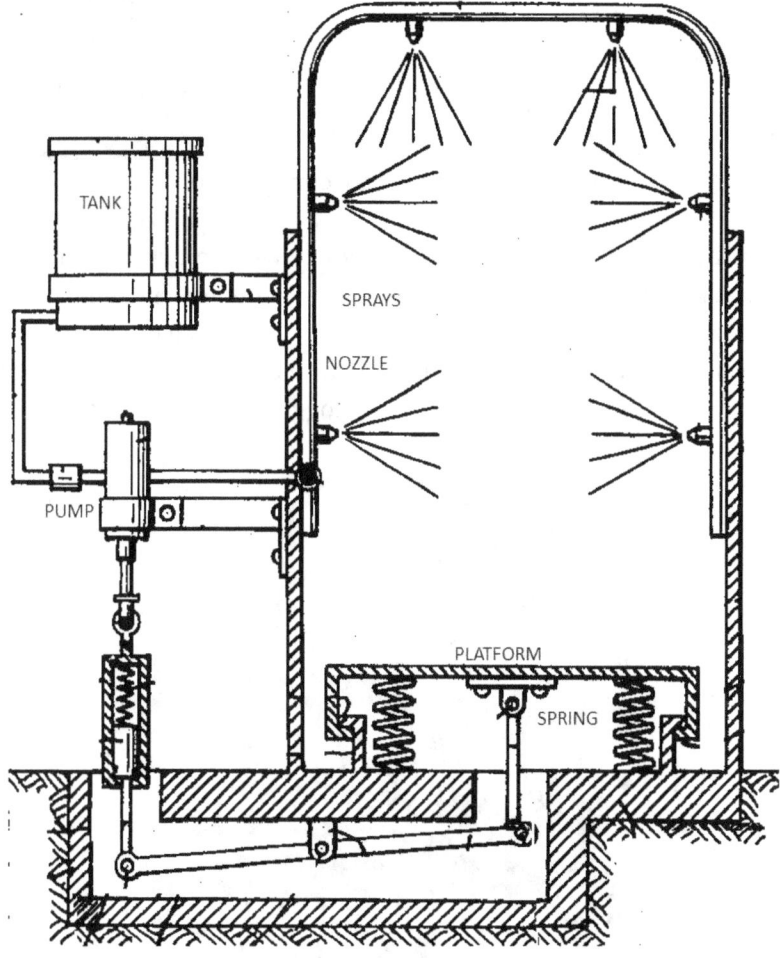

Figure 10-19: Sectional view of spray race (Cowan, 1971)

Power requirement for the pump can be supplied by a 6 to 8 horsepower stationary engine, a tractor power take-off, or a 5-horse-power electric motor. The discharged fluid collected on the floor of the tunnel and draining race is led to a sump and re-circulated. In addition to being cheaper to install than a dipping tank the spray race uses less liquid per animal and operates with a smaller quantity of wash, which can be freshly made up each day.

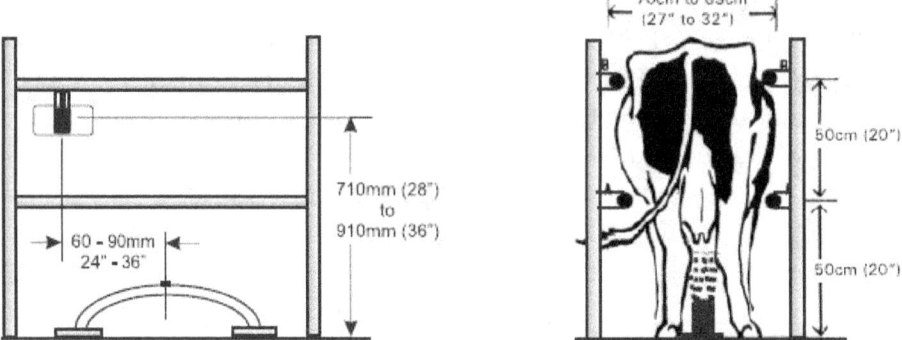

Figure 10-20: Spray race configuration

Comparing dipping and spraying methods, spraying is quicker than dipping and causes fewer disturbances to the animals. However, spray may not efficiently reach all parts of the body or penetrate a fur of long hair. The mechanical equipment used requires power, thus requires regular maintenance and replacement of worn out or damaged parts and the nozzles tend to get clogged and damaged by horns more easily.

Figure 10-21: Typical mobile spray race

Further reading

Bello R. S. (2011). *Combating Risks and Wastes in Cattle Examination and Drug-Use* 161-166 Ilorin, Kwara State, Proceedings of 42nd annual general meeting of the Nigerian Institution of Agricultural Engineers

Burns R. T (2006). Animal Waste Anaerobic Digester Basics Agricultural & Biosystems Engineering Department Iowa State University

Devendra C., McLeroy G.B., Goat and Sheep Production in the Tropics, Intermediate Tropical Agriculture Series, London, Longman Group Ltd., 1982.

Douglas W. Williams (2006). Dairy Joseph Gallo Farms Dairy Manure and Digester Manure Digesterdoug.williams@valleyairsolutions.com

Hafez E. S. E., 1975. The Behavior of Domestic Animals, 3rd Edn, Baltimore, The Williams and Wilkins Co.

Hall H. T. B. (1986). <u>Diseases and parasites of livestock in the tropics</u>. 2nd edition. Longman scientific & technical Pub. England.

Hall J. M., Sansoucy R., Open Yard Housing for Young Cattle, FAO Animal Production and Health Paper, no. 16, Rome, Food and Agriculture Organization of the United Nations, 1981.

Lennart P. Bengtsson and James H. Whitaker (Ed) (1988). *Farm structures in tropical climates* FAO/SIDA cooperative programme. Rural structures in east and south-east Africa Food and Agriculture Organization of the United Nations.

Turton J A. Methods of Tick control in cattle. www.nda.agric.za/publications date accessed 17/09/2012

Williamson G., Payne W.J.A., An Introduction to Animal Husbandry in the Tropics, 3rd ed., Tropical Agriculture Series, London, Longman Group Ltd., 1978.for meat production, but the by-products, such as pigskin, bristles and manure are also of economic

Test Questions

Question 1

   a. Briefly explain the factors to be considered in building farm house
   b. List the four critical factors that are required in successful farmstead planning and development.

Question 2

   a. List the components of the four zones in farmstead planning. Draw a layout diagram of the planning zones
   b. What are the factors affecting farmstead planning. Briefly explain them

Question 3

   a. List various means of farm transport and their description
   b. What are the two categories of farm transports? What are the purposes of these transport systems?

Question 4

   a. What are the four methods of planting seed? And their mode of operations
   b. What are the traditional functions of a seeding machine /planting equipment

Question 5

   a. Describe the process of moisture removal from agricultural products
   b. What is artificial drying? Give two examples.
   c. List two drying problems and their effects on agricultural material

Question 6

   a. What is natural drying and in which ways can this occur?
   b. What is artificial drying? Give two examples.
   c. List two drying problems and their effects

Question 7

Describe the following storage methods

    a. Traditional structures for tuber crops
    b. Bag storage
    c. Underground/pits storage
    d. Storage bins
    e. Improved traditional bins
    f. Cribs

Question 8

    a. What are the engineering problems involved in the soil and water conservation?
    b. List and explain the classes of soil water
    c. What are the various sources of irrigation water

Question 9

    a. Describe the following terms
    b. Water requirement
    c. Net irrigation requirement
    d. Gross irrigation requirement
    e. Irrigation efficiency

Question 10

    a. List and describe the three types of irrigation systems
    b. Compare the advantages and disadvantages of sprinkler irrigation and surface irrigation

Question 11

    a. What are the major components of a sprinkler system? Describe their functions
    b. Classify the sprinkler system and briefly describe each

Notes

Titles in author's list

Agriculture & mechanization series

- ♣ **Farm power and machinery operations**
- ♣ Agricultural machinery & mechanization
- ♣ Agricultural engineering: principles and practice (Vol 1)
- ♣ Agricultural engineering: principles and practice (Vol 2)
- ♣ Farm tractor systems: operations and maintenance
- ♣ Timeline of agricultural mechanization

Horticultural series

- ♣ Horticultural machinery: equipment and safety
- ♣ Fruits and vegetable technologies: management options

Workplace safety and machine technology series

- ♣ **Agricultural machinery hazards & safety practices**
- ♣ **Workplace hazards risks & control**
- ♣ Workshop technology & practice
- ♣ Technical drawing presentation and practice

Students' handbook series

- ♣ Study companion
- ♣ Path to exam success

Edited works on sustainable agriculture and environment series

- ♣ Sustainable agriculture: prospects and challenges
- ♣ Sustainable environmental management: issues and projections

More information available @:

http://www.amazon.com/Segun-R.-Bello/e/B008AL6RI0

www.ingramcontent.com/pod-product-compliance
Lightning Source LLC
Chambersburg PA
CBHW081104170526

45165CB00008B/2315